TOUGH PLANTS

First published in 2025 by OH
An Imprint of HEADLINE PUBLISHING GROUP

1

Cataloguing in Publication Data is available from the British Library

Hardback ISBN 978-1-03541-759-9

Printed and bound in China

Headline's policy is to use papers that are natural, renewable and recyclable
products and made from wood grown in well-managed forests and other
controlled sources. The logging and manufacturing processes are expected
to conform to the environmental regulations of the country of origin.

HEADLINE PUBLISHING GROUP
An Hachette UK Company
Carmelite House
50 Victoria Embankment
London EC4Y 0DZ

www.headline.co.uk
www.hachette.co.uk

TOUGH PLANTS

GARDEN GLADIATORS THAT PACK A PUNCH IN EXTREME WEATHER

VAL BOURNE

OH

CONTENTS

INTRODUCTION

CHANGING CLIMATE ACROSS THE GLOBE

Gardening is a great skill and it's so much more than simple physical activity, for gardeners follow a yearly rhythm that's dictated by the seasons and that's true across the whole planet.

As a result, experienced gardeners, wherever they are, sense when to plant or sow and they develop an empathy with the natural world. They are in tune with plants, animals and climate, more so than almost any other group of people. However, climate change is putting a spanner in the works. Many of the plants we used to rely upon no longer perform for us due to the erratic mixture of heavy rainfall and periods of drought. This book is all about the toughies: the resilient ones that continue to please.

RIGHT
Clematis
'Purpurea Plena
Elegans'.

6

MY OWN PATCH IN THE COTSWOLDS

We're all in the same boat. We're all wrestling with man-made climate change, because the world is warming up degree by degree. This is producing erratic weather conditions all around the globe and the seasonal weather patterns that we used to rely on have largely disappeared. Your winter may well be getting warmer and wetter, and that means that plants don't bother to produce their own form of anti-freeze. When they experience a sudden cold weather event, they are not prepared for it. We're all losing more plants than we did in the past and common 'cold-shock' casualties include penstemons, salvias, echinaceas and many silver-leafed plants.

Perpetually damp soil, lashed by heavy winter rain, spells sudden death for many smaller bulbs, especially dwarf narcissi. 'Cedric Morris', which used to be a midwinter marvel in my own garden in the Cotswolds, has become harder to keep. It's dwindled away to almost nothing. Many a plant fails to get a long-enough cool period of dormancy. Echinaceas struggle for me, because they hate stop-start winters. Others hate having wet roots. Even those that shrug off wet conditions can grow poorly, because nutrients continually get leached out of the soil.

Our springs are often drier than they used to be, with clearer skies. This means that days are warmer and sunnier, but clear nighttime skies see temperatures dip and this can lead to a 20°C (70°F) gap between day and night. This inhibits germination, because lower nighttime temperatures are too cold. Plants and seeds need nighttime temperatures of 10°C (50°F) upwards before they whoosh into growth. They hate extremes. They also need moist conditions when spring arrives, not dry soil. Gardeners, growers and farmers across the world are finding that the spring growth spurt they used to be able to rely on just isn't happening anymore.

Summer comes earlier than it did in many parts of the world. That creates problems for the natural world, as well as us. Plants react to the changes more readily than insects and animals do. The buddleia that used to sustain Vanessid butterflies in early August in my own garden flowers four weeks too early, so there are no small tortoiseshells, peacocks, red admirals or painted ladies drinking in the honey-scented nectar any more. This is a recent change, for there used be sixty or seventy butterflies on my *Buddleja davidii* in the 2000s. I've gone over to a later-flowering *B. x weyeriana* hybrid named 'Pink Pagoda' in an attempt to provide nectar when these butterflies are active. We used to get small coppers on our low-growing *Origanum vulgare* 'Compactum' every July. Now they're out of sync as well. And I could go on.

Summer droughts are prevalent too, right across the world, and we're inundated with books on drought-tolerant plants but they don't appreciate winter wet. If, like me, you're a green gardener trying to be sustainable, you face a conundrum. Saved rainwater only goes so far. After that, do you use up precious water on the borders when it's in short supply in the tap, or let the plants shrivel? I prefer not to use mains water, but that means the cultivars and varieties you once relied upon as bulletproof die back just when they should be spectacular. Disappointments sometimes include the lemon-yellow daisy, *Helianthus* 'Lemon Queen', and its favoured companion, *Eupatorium maculatum* (Atropurpureum Group). Both of these previous stalwarts are now in the lap of the gods in my garden. They fail in dry years and leave a maddening gap, as do most border phloxes. You hover over them with a spade in autumn, in 'do I or don't I?' mode. Shall I dig them up or not? Gardeners across the world are having the same dilemmas.

With these problems in mind, I have begun to create shade in the borders. A purple-leafed *Cotinus coggygria* 'Royal Purple' lords it over one part of the autumn border and prevents the perennials nearby from flagging in hot summer sun. It leafs up late and the leaves drop late too, so the wine-coloured foliage is particularly useful at highlighting beige-tinted tall late-season grasses. The roots also drain the soil and this helps the snowdrops planted under the canopy.

A 'viticella' clematis screen, consisting of 'Little Bas', 'Étoile Violette', 'Betty Corning' and 'Purpurea Plena Elegans', spans a series of tripods, shading another part of my autumn border. Casting shade is going to become more important than ever if these dry summers persist. One of the problems of climate change is the variability in weather patterns from year to year.

Autumn lingers longer these days. When I was a child in the 1950s, the first frost always cut back frost-tender plants, such as dahlias and runner beans, in the first half of September, just when they were at their best. These days I'm able to pick both in November, more often than not, and I might well see a foraging hedgehog on the way. Years ago, hedgehogs would have been tucked up weeks before, in their hibernacula. Now they can be seen in December, feasting on small slugs and grubs.

Balmy autumns create gardening problems. The leaves fall a month later than they did and the oaks around my cottage are often still in leaf in December. By the time you're able to rake up the fallen leaves, the bulbs and hellebore buds are already through! The warming climate also produces windier weather and that's challenging to plants, for wind is enemy number one when it comes to growth. It will also topple trees, especially if they're still in leaf. Worse still, the weeds keep growing.

Gardeners, growers and farmers are at the sharp end of climate change, so the plants I'm using are changing, not by my choice either! Many of my old favourites have fallen by the wayside, hence this book: *Tough Plants*.

PART 1

GARDEN STRUCTURE

FOR VISUAL EFFECT AND SHELTER

Garden structure is all about multi-level planting, using trees and shrubs as well as perennial plants.

I'm ashamed to say that I came to the realisation of its importance rather late in life. It happened after I moved to Spring Cottage in November 2005. I arrived with hellebores and spring bulbs that I had agreed to take from my old garden. They were duly planted, straight away, and they flowered without missing a beat. However, they looked like a cheap, gaudy bedding scheme.

Nothing wrong with that in a public park, but they were a visual disaster in a country garden. It took me some time to realise why. I began to see, somewhat slowly I have to say, that my woodlanders and spring bulbs needed a woody framework in order to shine, because most early flowering plants hug the ground. When I planted my first tree, a mere 1.2m-high (4ft) parasol, it all seemed to slot together. Note to self: woody structure should go in first, and not second, Val.

That first tree, by the way, is still a small parasol some twenty years on – slightly taller perhaps, but no fuller. It is an autumn-flowering cherry, *Prunus x subhirtella* 'Autumnalis', and it obviously resents the cold winters and iffy summers that my garden dishes up even more than I do. In fact, I deserve to be tried and sentenced for the crime of cruelty to plants, because this poor specimen hangs by a thread. Not dying, but not thriving either, my limbo-land autumn-flowering cherry.

I wanted one for nostalgia's sake, which is never a good reason to plant anything. You see, I had chased the blossom around the school playground on November days, because several thrived in gardens surrounding the grey asphalt yard tucked under the flight path of Heathrow. Chasing it around spared me the trouble of talking to other children, because I was a

RIGHT
Courtyard with box borders and balls, dahlias, *Salvia patens* 'Cambridge Blue' and *phlomis*.

OVERLEAF
Prunus 'Kursar' above spring planting, including hybrid hellebores.

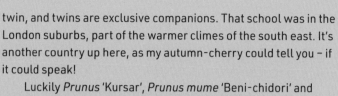

twin, and twins are exclusive companions. That school was in the London suburbs, part of the warmer climes of the south east. It's another country up here, as my autumn-cherry could tell you – if it could speak!

Luckily *Prunus* 'Kursar', *Prunus mume* 'Beni-chidori' and *Prunus incisa* 'Kojo-no-mai' have all done much better for me. They've been joined by fourteen witch hazels (*Hamamelis x intermedia*) and they provide resilient marmalade strands of flower every January without fail. Better still, the flowers don't brown in hard winter weather. These shrubby deciduous beauties have a branching shape that allows you to plant underneath them, right up to the trunk. They have been joined by several named forms of *Daphne bholua*, wintersweet, winter honeysuckle and viburnums (see Fragrance – p26 – for more information).

Woody plants offer more than perspective and fragrance, however. They provide overhead shelter and underground drainage. They shade and protect the ground, even when the branches are bare, and this overhead shelterbelt encourages plants to flower earlier than they would in bare ground. The woody root system helps to drain the soil and this is more important now that winters are milder and wetter. Warmer, drier ground means fewer losses as well as earlier flowers.

I plant densely and aim to cover the soil with low-level flowering plants by April, which isn't difficult for a self-confessed 'plantaholic' like me. Dense planting is a good thing. It keeps moisture in the ground, traps carbon into the soil and allows all sorts of organisms and creatures to thrive under the jungle-like layer. Ground beetles, the best predators of slugs and their eggs, thrive in these conditions. Early flowers sustain early pollinators, so they are important for the natural world, but every gardener should aim to have flowering plants for as much of the year as possible.

I've saved the best for last, though. Wherever you live in the world, the sun moves through the sky from dawn to dusk. An overhead canopy of boughs and branches allows light to filter through, creating a magic-lantern pattern of light and shade that changes throughout the day. It's like having a magician at your personal disposal!

FOLIAGE, FORM AND FERN FANATICS

When gardeners begin to make a garden, they always concentrate on flowers and often end up with a gaudy dolly mixture of colours.

I know. I've been there. It doesn't look good and it doesn't last very long either. Gradually the penny drops: foliage is really rather beautiful and it pleases the eye for far longer. It should be a key element in every garden. It acts as a buffer, separating and supporting the colourful but short-lived divas we all fall for. You can repeat it through a border, thereby uniting the planting scheme. I have become a fern fanatic purely because I love the architectural nature of the fronds, and these primitive beings don't even have any flowers!

Don't panic though. Lots of flowering plants have interesting foliage as well. I use two plants in a sunny border, leading to the main cottage door, and both have yellow flowers. One is a lucid-yellow daisy named *Anthemic tinctoria* 'E.C. Buxton', the other a dome-headed acrid-yellow achillea named 'Moonshine'. They overlap and run through the border like a spinning thread produced by Rumpelstiltskin.

Both share finely cut filigree foliage, although the anthemis is green-leafed and the achillea is silver. It's the foliage that ties them together, because it creates a texture that leads the eye up the path. It's important to repeat good plants, don't dot just one in or you'll get a pincushion. Train yourself to buy the same plant in threes, fives or sevens. Your garden is bound to improve.

READING THE FOLIAGE

It's important to learn how to 'read' the foliage, because it helps you to place plants into the correct position, the one where they happily thrive and prosper. Get it wrong and the odds are stacked against you before you start.

RIGHT
Trillium flexipes
x *sulcatum*
'Harvington
Dusky Pink'
hybrid and
Polystichum
setiferum under
witch hazels
in the author's
garden.

SILVERS FOR HOTSPOTS

Silvery foliage is often aromatic and its oily coating is a sunscreen designed to protect the foliage, because it's evolved in a sunny, hot part of the world – most probably the Mediterranean.

The winter, which is warm and wet, is the main growing season. The summers are hot and dry. If you cut Mediterranean plants back in autumn, in colder parts of the world, they will still start into growth. The fresh, vulnerable new shoots will die off as soon as cold weather sets in; aromatic plants need to be left intact during winter and they can look shabby. When spring proper arrives, they can be trimmed back hard. Borderline tender plants, like fuchsias, salvias and penstemons, are all treated in the same way.

Aromatic plants are perfect for hotter, drier spots because the pale, silver foliage doesn't absorb very much heat from the sun. The sparkling foliage is an excellent foil for blues and purples. Many of these silvery plants manage to survive wet, mild winters and their foliage can be a star attraction because the low sunlight picks up the

detail. *Phlomis italica*, for instance, is a woody subshrub with rigid stems clothed in ovate silver foliage. When winter sun strikes, the foliage looks almost quilted and it's one of those tactile plants you can't resist stroking.

Silver-leafed plants are drought-tolerant, but only once established. They rely on deep root systems that can penetrate to a depth of 1.5m (5ft) or more to seek out moisture and nutrients. This root system takes at least one growing season to develop. These plants need nurturing during their first growing season, because their roots have been constricted by a flower pot. The words 'drought tolerant' will only apply in subsequent years.

The flowers of aromatic plants produce concentrated nectar, so they are bee and butterfly magnets in summertime. The most sugar-packed nectar, containing 78 per cent sugar, belongs to marjoram, or *Origanum vulgare*. The flowerheads are a mixture of tiny pink and purple florets. Marjoram attracts bees, but so do salvias, lavenders and catmints.

GREEN AND GLORIOUS

Green foliage is the norm when it comes to colour and it's also the most versatile when it comes to growing conditions.

A bright garden position is generally all that's required. Deciduous plants, ones that shed their foliage or retreat underground, are hardier than evergreens. They don't shut down for a period of dormancy, because they shed foliage throughout the year. Sometimes evergreens shed their leaves after severe weather, but they usually survive. The foliage on *Daphne bholua* is evergreen in most gardens but disappears in my cold garden, although this beautiful winter-flowering daphne has persisted from year to year.

Green foliage can be soft and shiny and this should tell you that this plant will be happiest in sheltered semi-shade because the foliage doesn't have a tough, leathery coating. If it is leathery, it will be tough enough to withstand winter, although it probably became leathery in order to deter grazing animals, not because of cold.

Evergreens have the best foliage of all in winter light. It warms the cockles of your heart, whether it's box (*Buxus sempervirens*), or yew (*Taxus baccata*), or *Sarcococca confusa*, or a perennial such as *Epimedium* 'Spine Tingler'. The wave of green will enrich your garden and many evergreens are able to survive the vagaries of the topsy-turvy weather.

VARIEGATION – USE SPARINGLY, LIKE CHILLI SAUCE

Variegation should be used in moderation to cast light and shade among your rich greens. And I say this having seen gardens planted up with masses of variegation for no other reason than it's variegated. Some people – including a renowned nurseryman I know – are slaves to variegated plants! It's important to segregate foliage into cool and warm colours. Green and gold variegation is a warm mixture, although the colour can vary between custard and mayonnaise. This warmly tinted combination goes best with bright yellows and true-blues. You could emphasise the 'shaft of sunlight' theme with a spring-zing spurge and *Euphorbia epithymoides* is the most useful. The foot-high pouffe of acid-yellow arrives with the blue scillas and blue bulbs.

The cooler variegation of grey-green and ivory-white is far more subdued in tone. Like Hitchcock blondes, they're cool but tough. Equally cool-toned pinks and purples go well with this subtler foliage and it can tolerate a brighter position as well. Variegated foliage, usually caused by a genetic chimera, has leaves that are only partly green. The interesting thing is:

LEFT
Melica uniflora f. *albida*, *Adiantum aleuticum* 'Imbricatum' and other hardy ferns in a shady corner in the author's garden.

21

variegated plants tend to need less water. The variegated border phlox, 'Norah Leigh', for instance, does not flag in heat and dry conditions as so many border phloxes do. This was noticeable during the three-year Phlox Trial held at RHS Wisley, which finished in 2013. It stood out, somewhat defiantly, among the flagging masses, in drought and downpour. I have observed the same sort of resilience in other variegated plants.

GOLDEN GLIMMERS

Using golden foliage, on the other hand, is a balancing act. It can scorch in hot sunshine and the golden-leafed mock-orange *Philadelphus coronarius* 'Aureus' is a prime example. It needs tucking away in light, dappled shade. However, some yellows will not brighten and develop a golden patina unless they are in bright light. This applies to a golden hart's tongue fern, *Asplenium scolopendrium* 'Golden Queen', and an ivy named 'Buttercup'. They stay green in shade.

Very few plants manage to keep their golden foliage into summer and winter, either. This is why I value Bowles's golden grass (*Milium effusum* 'Aureum') and the golden great woodrush (*Luzula sylvatica* 'Aurea'). They provide yellow glimmers on bare soil and both glow in winter light. They, rather like yellow-splashed variegation, light up a dull winter's day.

SPLASHES AND FROSTING

You will also find plants with dappled and marked foliage and this indicates a woodland preference. Some dappling disappears as the plant flowers, and this happens with erythroniums such as 'Pagoda'. Arum lily (or *Arum italicum* subsp. *italicum* 'Marmoratum') on the other hand, has arrow-shaped leaves heavily and regularly decorated in winter white. The leaves are a feature from November onwards and they keep their colouration until they gently wither away in late spring. The autumn-flowering cyclamen, *C. hederifolium*, is another winter stunner, although there is considerable variation in the foliage. They can be all-silver, to mottled, to kaleidoscope arrangements of green and jade. I have spent many a happy hour browsing through trays of them in nurseries and garden centres, because there is always treasure to be found.

Pulmonarias and brunneras cast their own ice crystals with their frosted and splashed foliage. The tasteful *Brunnera macrophylla* 'Jack Frost' has been joined by many others, some with enormous leaves. However, I still prefer the dainty blue sprays of flower and the green-edged and veined silver foliage of this one. Some of them look as though they've been fed a diet of bodybuilder steroids, with leaves as large as giant hostas.

When it comes to pulmonarias, or lungworts, their ability to produce seedlings courtesy of the bees has provided a wide choice. I've found the square-spotted silver and green leaves and brick-red flowers of 'Leopard', a 1970s find from Graham Stuart Thomas's private garden, pleasing and persistent. 'Opal' has almost glacial-blue flowers and neatly spotted foliage and it produces masses of flower. Of the all-silvers, I find the narrow leaves of 'Samurai' and its cobalt-blue flowers form a stronger plant than the lovely violet-flowered 'Diana Clare'. One tip, don't cut the all-silvered leafed pulmonarias back hard after flowering. Give them a gentle tidy-up instead.

There are so many pulmonarias on offer and they slot into woodland borders really well. Their flowers open pink and turn blue once pollinated, and this has earnt them the common name of 'soldiers and sailors'. Strong-blues also feature and 'Blue Ensign' is a cracker, although the leaves are a rough-textured dark green. Many a woodlander's leaf has a red back and this is a light-trap system, indicating a need for dappled shade.

The message is clear: don't be seduced by flower alone. Scout the garden centres and nurseries for interesting foliage, because gardens need a chorus line as well as a celebrity line up.

FRAGRANCE

Fragrance is an important consideration, because it pleases the hedonist hidden deeply within every gardener.

The pleasure comes in so many ways, often unexpectedly, as it wafts through the air. The scent of our native honeysuckle, *Lonicera periclymenum*, is divine on a warm summer's evening. Shakespeare called it 'luscious' and the woodbine was and is still endemic in the lanes of Warwickshire, so he would have known. Whenever I catch the honeyed scent, it spins me back to a time when I travelled those lanes in an open-topped sportscar, a young and carefree teenager.

Then there is the rose that makes your nose fizz with a mixture of tea, fruit and lily of the valley. The coconut-infused rose scent of certain peonies at midday and the citrus waft from philadelphus blossom... they all make me go weak at the knees. As does the heady lily, too much for some, and the silvery-mauve moth-pleasers from sweet rocket to soapwort to single dianthus. Or perhaps you'd prefer to get overwhelmed with the honeyed buddleia growing in a warm spot.

Many fragrant lovelies can be captured in containers, like the hyacinth. Or they can be simply left to scramble into the light, as the woodbine does. The thing they all need though, before they agree to release any of their honey-trap scent, is shelter and warmth. This allows the nectar to flow.

Fragrance is not for our delight alone. It's designed to lure in the pollinator, so our pleasure is simply a byproduct – lovely though it is. It's used by paler flowers a great deal: the most-scented peonies are pastel pinks and creams, for instance. The lemon-scented and ivory-coloured 'Duchesse de Nemours' and the silver-edged pink bombe-shaped flowers of 'Mon. Jules Elie', both forms of *Paeonia lactiflora* raised in 19th-century France, were favourite cut flowers for the Parisian bourgeoisie.

Chinese *lactiflora* peonies combine fragrance and flouncy petals and they capture the summery softness of blue skies and cotton-

wool clouds. Winter would ravage them, so winter flowers need to be small and insignificant in order to defy the weather winter hands out. They lure their pollinators in with scent alone and it has to be a powerful formula because it needs to register in lower temperatures.

The best winter scent in my garden emanates from the aptly named wintersweet, *Chimonanthus praecox*, literally meaning 'early winter flower'. Although a Chinese native, one of the common names is Japanese allspice. This flowers on bare branches and the translucent watery-yellow flowers have a jolly touch of cranberry-red in the heart of each flower. In years past, this would have been given a sheltered site, because the flowers get browned by frost or they turn to mush in heavy rain. The rain is the most damaging of the two, I find. I grow wintersweet in the open, not against a warm wall. When it's on song, this large, rather untidy shrub is superb in winter in its Dr Jekyll incarnation. In summer, it's definitely Mr Hyde though, with large green leaves and pear-shaped fruits on a large, untidy shrub.

I forgive those summer shortcomings because wintersweet picks so well and, on a winter's day, the scent is sweet sugar and spice. It's also very good at providing woodland shade for all manner of plants, from trilliums, to hepaticas, to snowdrops and ferns.

The winter honeysuckle, *Lonicera x purpusii* 'Winter Beauty', is just as ungainly in summer as the wintersweet. However, it's another winter highlight with its early ivory-white tufts of flower. It feeds my neighbour's honeybees whenever January

BELOW
A honeybee on *Lonicera x purpusii* 'Winter Beauty' in early January in the author's garden.

produces a warmer afternoon. And again, it offers benign shelter for woodlanders and bulbs. I forgive both for their ungainly habit, because their scent is unsurpassed in the cold midwinter.

Hybrid witch hazels, *Hamamelis x intermedia*, offer a variety of scents from toilet cleaner to surgical spirit to highly floral, so I always advise people to buy them in flower and sniff them. The reds are generally unpleasant, to my untrained nose anyway, and many have the typical witch hazel smell, musty and sweet and (frankly) nothing to write home about. The best two for scent, similar to freesia, are the pale-lemon 'Pallida' and the butterscotch 'Aurora'.

Both produce lots of flower if they get deep soil and plenty of summer rainfall. These are, after all, interspecific hybrids involving Asian and North American species. Their larger flowers are more spectacular and their growth habit more vigorous, due to their hybrid nature. However, they are thirsty plants because they have one foot in the rainy season experienced in China and Japan. When they're in need of a drink, they slant their leaves to the vertical to avoid transpiration and this is a sign that they need water. No dribbling hoses please, carefully tip a bucket of water around each plant a couple of times a week. They will reward you with fistfuls of snuff-brown buds by the autumn.

Not everyone has room for witch hazels and large shrubby Jekyll and Hyde characters. If space only allows you a pot, or a balcony, I urge you to plant Christmas box, or *Sarcococca*. This smallish evergreen has small flowers, mere collections of stamens, set among evergreen foliage. The flowers pack a heady scent and every garden should have at least one close to a well-used path or gateway. The pink-toned flowers are more strongly scented and *S. hookeriana* var. *digyna* is a good choice, although the foliage develops a yellowish tinge if it gets starved in a pot. I prefer the truly green foliage and ivory-white flowers of *S. confusa*, although I don't find the scent quite as strong. Both produce black berries.

These early flowering fragrant shrubs offer vital cover and extra drainage for small bulbs and woodlanders that might otherwise be killed off by the heavy winter rainfall we seem to be getting thanks to climate change.

GRASSES FOR MOVEMENT AND TEXTURE

No other plant provides as much movement, from slow waltz to foxtrot to tango, as grasses and grass-like plants do.

They offer cadence and flow and, when low light descends, every detail gets the spotlight treatment. Many are forced to grow at the northern edges of their range, so they either flower late or not at all. My cold garden, in the heart of England, means that my named forms of *Miscanthus sinensis* don't produce their flowery awns until the beginning of September. Yet, in warmer spots they can flower in July. I recommend those with gardens like mine, in colder areas of the northern hemisphere, to grow a miscanthus called 'Silberfeder', because it's reliable and early. I fail with all panicums, American Switch grasses, even though they flourish in chilly American states. They need more summer warmth than cooler gardens can provide.

I adore pennisetums. However, many of the supposedly perennial ones, like *Pennisetum orientale*, don't survive my wet winters. One nursery, based in a maritime site in the drier eastern half of England, advertises it as fully hardy and perennial. Mutter, mutter! I have to treat it as an annual, so it can't be classified as tough. Still, I love the arching bottlebrushes produced by pennisetums, especially the ones that look like woolly bear caterpillars balancing on trapeze wires.

I do manage to grow 'Cassian's Choice' and this one is a chocolate-coated woolly bear. It's a form of *P. alopecuroides*, the Chinese fountain grass. This is the hardiest species of all, although they all need good drainage, space and a bright position. Many of them flower very late in colder gardens, or not at all. However, 'Karley Rose' bucks the trend and produces slender, pink fluffy heads in summer just as the echinacea flowers open. The grassy heads fade to beige in autumn and, when they're studded with tiny dewdrops on a misty October morning, I can't resist frisking them.

Certain grasses thrive and one of them is the fountain grass, *Stipa gigantea*. This is tall, 2m (6.5ft), with silvery fine foliage. As late spring turns to summer, stems rise up and splay themselves out like quills on a giant's pen. These unfurl with the first roses and they form a golden series of tiny triangles on quaking stems. They shimmer above alliums and early herbaceous, before fading in the autumn. It's a terrific unifier in a border because it acts as a shimmering backbone, forms a large clump and returns year after year.

Stipas are like cats in an armchair: they do not want to be disturbed, thank you very much. Should you get gung-ho and decide to lift and divide a stipa, most will turn up their toes in disgust. Other rugged stipas of note include *Stipa ichu*, a collection of wispy white plumes about 1m (3ft) in height. It waves at you in the slightest breeze and is an easier, although less spectacular, option than *Stipa barbata* – the real Rapunzel of grasses. This is tricky, but worth the struggle of sowing the sharp-tipped seeds, found at the opposite end of every ostrich feather, into coarse sand. Expect 50 per cent to grow.

My top shade-loving grass is *Melica uniflora* f. *albida*. The tiny, ivory-white rice-like beads shimmer above bright-green grassy leaves, like a fountain enclosed by a green ruff. This gentle grass, which does self-seed, can unify a shady border where ferns, hostas or woodlanders abound. It's a Chelsea Flower Show favourite, because it's early season.

Sedges usually thrive in damp soil, but the tightly waisted *Carex testacea*, the orange New Zealand sedge, is definitely a drought-resistant toughie. The fibre-optic fine tines look good in a pot, or on the ground because they swirl around like a vortex. The orange colouring makes it stand out and this is a striking specimen plant with a year-round presence. Narrow-leafed carexes, usually best in drier soil, also come in jade-green and cola-brown. The broader-leafed ones need to be cut back by half, and only *half*, during the growing season. Don't do it in autumn or winter! I've committed both crimes – hard pruning at the wrong time – and they've died!

Late-season grasses include *Miscanthus sinensis* and pampas grass, *Cortaderia selloana*, and both have grassy heads that endure over winter. The South American pampas grass has razor-sharp leaves, so place it carefully.

PART 2

THE
PLANTS

'ROBINSONIANA'

ANEMONE NEMOROSA

**Gently nodding grey-blue flowers, with thunder-cloud backs,
opening a week or so before the leafy canopy unfurls.**

Wood anemone flowers only open wide, to allow pollinators
to work their magic, when the skies brighten. On dull days, the
flowers stay tightly shut. They form carpets over time, spreading
faster in moister soil, so they're great gap fillers between bulbs
and ferns. These summer-dormant woodlanders don't enjoy dry
conditions, so wetter winters suit them well.

They're found growing naturally in the woods of Europe, and
colours vary according to soil and location. 'Robinsoniana' came

PLANTING ADVICE
Give them moist shade close to deciduous trees and shrubs. Divide just as they break into growth in spring. The blues are normally stronger.

from Ireland, home to many wild blue wood anemones. It was spotted at the Oxford Botanic Garden in 1870, possibly by the famous gardener William Robinson. The large blue-grey flowers appear a little later than many others, so this one extends the season. Pink wood anemones grow in the woods of Kent. 'Westwell Pink', found near Ashford, was distributed by local nurseryman Tim Ingram.

Anemone ranunculoides is less desirable because the bright-yellow, buttercup-sized flowers lack grace and charm. It spreads too readily and the foliage is a dull green. There are double forms, including 'Ferguson's Fancy' and 'Pleniflora'. However, it's a parent, along with *A. nemorosa*, to a stunning pallid-yellow hybrid named *Anemone x lipsiensis*. This choice wood anemone casts a shaft of gentle moonlight on bare spring earth. Grow it!
Height 10cm (4in) / H4–8

FIVE MORE WOOD ANEMONES TO TRY

'VESTAL'
Late to flower, with fully double neat-as-neat white rosettes. It makes an excellent foil planted among sooty hellebores and it also lights up shade. Available as a dry bulb. Height 10cm (4in) / H4–8

'ALLENII'
Large flowers in lavender-blue. Raised by James Allen before 1901 and possibly wild-collected in Northamptonshire, where Allen had his nursery. Height 10cm (4in) / H4–8

'VIRESCENS'
Leafy bracts in pale-green, standing out against the bare earth like intricate doilies, but sadly no flower. Height 10cm (4in) / H4–8

'BOWLES PURPLE'
Lilac flowers that darken after opening, this has fuller, rounder petals with reddish-purple backs. Height 10cm (4in) / H4–8

ANEMONE APENNINA
From the Apennine Mountains, although it seems to thrive in similar conditions to *A. nemorosa*. Softer lavender-blue flowers are held above finer, more divided foliage. Height 10cm (4in) / H4–8

'JACK FROST'

BRUNNERA MACROPHYLLA

Silvered heart-shaped leaves, rimmed and veined in cool-green, framing airy wands of tiny, forget-me-not blue flowers.

Brunnera macrophylla hails from Siberia and its common name is Siberian bugloss, so this shade lover is healthy, tough, very hardy and resilient. In 2000, Terra Nova Nurseries introduced the silver-leafed 'Jack Frost', one of the finest plants to emerge in recent times. It's said to be a sport of 'Langtrees', one of the named green-leafed selections, and it was originally bulked up by micropropagation. However, 'Jack Frost' will divide in early spring, just as it begins to shoot, although there's no need to do so routinely because it isn't an aggressive grower.

FOUR MORE SPRING FOLIAGE PLANTS TO TRY

ATHYRIUM NIPPONICUM VAR. PICTUM 'SILVER FALLS'

Japanese painted ferns produce silver-washed fronds with pink-purple highlights. This 2000 introduction from Oregon's Diana Ballantyne unfurls fabulous fronds in late spring. Height 40cm (16in) H4–9

EPIMEDIUM 'PINK CHAMPAGNE'

The new spring foliage, typically green and heart-shaped, is heavily splashed in rhubarb-pink and burgundy. Pink 'dangling-spider' flowers follow, trembling on wiry stems. Raised by Darrell Probst in America. Height 40cm (16in) H5–9

HEPATICA NOBILIS 'STAINED GLASS'

From John Massey's Ashwood Nursery in the UK, this hepatica combines blue flowers with highly decorative tri-lobed foliage in jade-green over-washed in pink. Dark leaf margins and veins resemble leaded glass. Height 15cm (6in) / H4–8

HEUCHERA 'GREEN SPICE'

Choosing one heuchera from the multitude is almost an impossibility. This award-winning heuchera has frosted green-washed foliage brightened by red veins. Height 30cm (12in) / H4–9

More silver-leafed brunneras have appeared since, including the white-flowered 'Mr Morse', raised in Belgium by Chris Ghyselen. Others, such as 'Alexander's Giant', have enormous leaves that dwarf most spring-flowering, shade-loving woodlanders. You can also find 'Silver Heart' and 'Looking Glass'. The latter is highly regarded.

Variegated brunneras are also available but are not as robust. 'Hadspen Cream' was raised by Eric Smith in the 1970s in the days when Penelope Hobhouse lived at Hadspen House in Somerset. Height 45cm (18 in) H3–9

PLANTING ADVICE
Never cover the crown and this is good advice for most woodlanders. A general slow-release pelleted fertiliser can be applied in early spring.

HYBRID HELLEBORE

HELLEBORUS X HYBRIDUS

Long-lasting rugged tepals, rather than weather-prone petals, make this long-lived bee-pleaser a must.

There are between seventeen and twenty species of hellebore in the wild, depending on which botanist you believe. They have provided a genetic melting pot of colour, foliage and form and this has allowed plant breeders to produce a varied range of hybrid hellebores over the last fifty years. The colours range from clear-white through to slate-black. Some have picotee edging, others have vivid nectaries, some are heavily spotted and there are also doubles and anemone-centred forms too. Look for good foliage and perfectly formed flowers.

Bear in mind that darker purples, wine-reds and slate-blacks disappear visually unless they are placed close to a bright bolt of colour. This might be a ghostly birch trunk, blue bulbs or

FOUR MORE HELLEBORES TO TRY

HELLEBORUS FOETIDUS
This British native does well in dank shade and the cluster of small bright-green, red-edged green flowers glows. Height 35cm (14in) / H6–9

HELLEBORUS 'PENNY'S PINK'
One of the best of Rodney Davey's sterile hybrids, with marbled foliage and long-lasting rose-madder flowers that darken with age. Height 40cm (16in) / H4–8

HELLEBORUS NIGER - THE CHRISTMAS ROSE
Best treated as a pot plant to prevent the early white flowers getting muddied, but worth growing for its pristine midwinter flowers. Height 25cm (10in) / H4–8

HELLEBORUS 'WALBERTON'S ROSEMARY'
Raised by the late David Tristram of Walberton's Nursery, this sterile hybrid between *H. niger* and *H. x hybridus* produces outward-facing, dusky-pink open flowers. Height 35cm (14in) / H4–8

rich evergreen foliage. Major on the paler colours – the apple-blossom pinks, whites, pale-greens and pallid-yellows – if you want to make an impact in the spring border.

Pot-grown hellebores can be root-bound, so don't be afraid to trim any circling roots off and do soak the plant well. Be careful to position the plant so that it's level with the soil surface. Plant when the weather's clement and water well during the first growing season. All established hellebores should have their foliage removed in early December, because they suffer from a leaf-spot disease called *Microsphaeropsis hellebori* (syn. *Coniothyrium hellebori*). This also shows the flowers to best effect. Leave the foliage on smaller plants. Height 35cm (14in) / H6–8

PLANTING ADVICE
Hybrid, also called oriental hellebores, prefer dappled shade and good soil that doesn't get waterlogged in winter. They are greedy feeders, so apply a high-potash fertiliser in early spring and again straight after flowering. Deadheading is advised, because hybrid seedlings will vary.

EUPHORBIA EPITHYMOIDES

A foot-high mound of long-lasting acid-yellow – to brighten the dullest spring day.

Euphorbias, or spurges, provide a long-lasting presence because their tiny flowers are surrounded by rugged leafy bracts. These can come in spring-zing green, acid-yellow or burnt orange and they can last for months. The red bracts of the poinsettia, *Euphorbia pulcherrima*, indicate the range and diversity within the Euphorbiaceae, the world's fifth-largest flowering plant family. Many are evergreen or tender, but *E. epithymoides* (previously

FIVE MORE EUPHORBIAS FOR SPRING ZING TO TRY

EUPHORBIA GRIFFITHII 'FIREGLOW'

This rambling Himalayan spurge produces asparagus-like spears in April. They produce olive-green foliage, mid-ribbed in white, topped by orange bracts. Height 90cm (3ft) / H6–9

EUPHORBIA MYRSINITES

Succulent grey foliage, indicating a need for good drainage and sun, is topped by a small cluster of acid-yellow flowers in late spring. Good on a scree or in a raised bed. Height 15cm (6in) / H5–9

EUPHORBIA AMYGDALOIDES 'PURPUREA'

A British native evergreen spurge, found in deep shade. This purple-leafed form has lime-green heads of flower. Height 35cm (14in) / H4–9

EUPHORBIA CHARACIAS

The classic Mediterranean plant of dry slopes, found from Portugal to Morocco through to western Turkey, with almost furry, pale-grey leaves topped by a crook of yellow flowers. Find a hot spot. Height 1m (39in) / H7–9

EUPHORBIA 'MINER'S MERLOT'

Lots of compact hybrid euphorbias with wine-red, evergreen leaves have appeared with names like 'Blackbird' and 'Redstart'. This one has wine-coloured foliage and bright lime-yellow flowers, bridging spring and summer. Height 35cm (14in) / H4–9

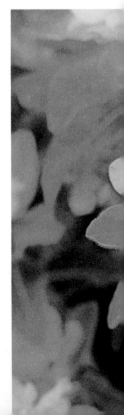

Lengthen the lifespan by
cutting *E. epithymoides*
back to nothing in mid-
May, to encourage lots
of new growth at the
base. Use these fresh
shoots for cuttings. If
you wish to divide, do so
in early spring.

called *E. polychroma*) is a deciduous, truly hardy
euphorbia that's native to cooler upland areas of Europe.
It even survives in Finnish gardens and in colder parts of
the USA.

The cushion spurge is able to flower in shade, despite
a preference for brighter dappled shade, and it's the
perfect foil for early spring miniature blue muscari (grape
hyacinth) and scilla. It persists through April and May and
the acid-yellow pouffe, reaching a mere foot in height,
could underpin dark tulips, such as 'Paul Scherer' and
'Queen of Night'.

E. epithymoides also makes a good container plant
and Japanese Hakon Grass (*Hakonechloa macra*) is a
good partner. 'Nicolas' produces new green leaves in
April and the foliage curtsies over the edge of the pot,
contrasting with the acid-yellow domes of the euphorbia.
Height 30cm (1 ft) / H4–8

'TREVI FOUNTAIN'

PULMONARIA

A fountain of heavily spotted linear leaves, topped with cobalt-blue flowers – reminiscent of a summer sky over Rome.

Pulmonarias should be an essential part of anyone's early shady border, because they provide clusters of pollinator-friendly flowers by mid-spring, just when flowers are few and far between. The flowers open pink but turn a shade of blue once pollinated, and this two-tone look gave rise to the name of 'soldiers and sailors'. Like many members of the borage family, they replenish their nectar quickly. Better still, the heavily spotted foliage endures over winter, although it may need tidying after hard weather.

'Trevi Fountain', which has *P. longifolia* in its bloodline, was raised by Terra Nova Nurseries c. 1999 and it's proved a huge success, even in the southern states of the USA. It holds numerous

FOUR MORE PULMONARIAS TO TRY

'OPAL' ('OCUPOL')
The almost glacial-grey flowers, which open pale-pink, show up well against the bare ground and the foliage is also stunning. Height 25cm (10in) / H3–9

'BLUE ENSIGN'
A seedling in Bowles's Corner at RHS Wisley, this green-leafed pulmonaria has the richest blue flowers of all. Needs regular division. Height 25cm (10in) / H3–9

'VICTORIAN BROOCH'
Also from Terra Nova, in 1998, and rated by Dan Heims as his finest. The blue and red flowers, along with the more rounded, silver-freckled green foliage, make this compact lungwort sing out. Height 25cm (10in) / H3–9

'LEOPARD'
Found in the garden of Graham Stuart Thomas in 1977, the green foliage should have dice-shaped, almost square, splashes regularly spaced. The flowers are warm-red. Height 25cm (10in) / H3–9

PLANTING ADVICE
Dappled shade and good
soil are ideal and do plant
them quickly: they
resent being in a pot.
Water thoroughly in the
first season and divide
them every four years,
discarding any woody
stems.

showy flowers above the equally impressive foliage and that has
made me single this stunner out among the many. It's strong and
resilient and not all pulmonarias are. I, like many, have struggled
with the dainty 'Sissinghurst White' and the variegated red-
flowered 'David Ward'. I have lost 'Roy Davidson' and 'Mrs Kittle'
and I have abandoned *Pulmonaria rubra*, despite its early brick-
red flowers, because it tends to cover too much ground.

It's always worth taking time and trouble to deadhead
pulmonarias after flowering, to prevent inferior seedlings. Most
can be sheared back after flowering, with the exception of the
violet-flowered silver-leafed 'Diana Clare'. Named after Bob
Brown's wife, this needs a gentler treatment.
Height 25cm (10in) / H3–9

THE PLANTS

'RICHARD KAYSE'

POLYPODIUM CAMBRICUM

'Richard Kayse' produces almost 3-D fronds, like a series of evergreen fir trees in miniature.

Plants with enduring winter foliage are as rare as hen's teeth, but this fern has a secret. It lies dormant over summer and produces new fronds in late August or early September. Consequently, it looks pristine just when everything else is either underground, or in decline. It's as if J.R.R. Tolkien had left one of his Christmas letters to his children. *Polypodium* means 'many-footed', so this fern spreads to form wide clumps but leaves convenient gaps that allow small bulbs, such as snowdrops and scillas, to wander through.

PLANTING ADVICE

These easily grown ferns enjoy growing in the lee of spring-flowering deciduous shrubs, whether it's viburnums or witch hazels. Polypodies do need some lime, so they may not do for those on acidic soil.

There are many forms but 'Richard Kayse' is arguably the finest and one of the oldest in cultivation. It was collected in 1668 from a sheer rock face in Dinas Powys by the aforementioned Richard Kayse, a Bristolian. It was reintroduced into horticulture by Martin Rickard, a British fern expert. This sterile fern does not produce spores, but it's easily divided in late summer when dormant – although polypodies do not need rouline division. Height 35cm (14in) / H5–8

FOUR MORE WINTER FOLIAGE FERNS TO TRY

POLYSTICHUM SETIFERUM 'PULCHERRIMUM BEVIS'

The best large and luxuriant upright soft shield fern with intricate fronds that taper to a slightly curved tip. Found in a hedgerow near Axminster in 1876, by a gentleman called Bevis. Height 1m (39in) / H5–8

POLYPODIUM CAMBRICUM 'PULCHERRIMUM ADDISON'

A polypody with large deep-green fronds, divided and layered so that it looks lacy. Height 35cm (14in) / H5–8

POLYPODIUM X MANTONIAE 'CORNUBIENSE'

Bright-green highly divided fronds that gracefully arch over. Height 35cm (14in) / H5–8

ASPLENIUM SCOLOPENDRIUM - THE HART'S TONGUE FERN

Keep this evergreen fern away from extreme conditions, whether it be strong sunshine or cold temperatures. The solid linear leaves produce variations. Height 35cm (14in) / H4–9

'GREEN SPICE'

HEUCHERA

An exquisite heuchera, with lobed dark edging framing silver foliage, heavily veined in wine-red.

Another great shade-tolerant plant from Terra Nova Nurseries. Although launched in 1993, it's still one of the most widely grown heucheras some thirty years later. The silvery texture shows up in winter light and mixes far more easily with the spring palette than the dark-leafed heucheras, yellows, limes and oranges. The foliage of 'Green Spice' loses some sparkle in summer, but turns 'neon-pumpkin orange' in autumn, according to Dan Heims.

Heucheras benefit from regular division. It keeps plants vigorous, lengthens their lifespan and prevents the fleshy stems

FOUR MORE HEUCHERAS FOR WINTER EFFECT IN SHADE TO TRY

'BLACKBERRY JAM'

Deep blackberry-maroon foliage with charcoal veins that have a silver shimmer. Stays compact. Height 25cm (10in) / H4-9

'BLONDIE IN LIME' (LITTLE CUTIE SERIES)

A petite heuchera with limey-green foliage topped with heads of lemony-green flowers in spring, summer and autumn. Height 25cm (10in) / H4-9

'GLITTER'

Silver foliage etched with black veins and colour-washed in violet. Fuchsia pink flowers throughout spring and summer so this one has flower power too. Height 25cm (10in) / H4-9

'SILVER CELEBRATION'

Raised by Plantagogo and named to celebrate this British heuchera specialist's 25th birthday. The large metallic pewter leaves have a dark-red underside, indicating a need for shade, and large white flowers follow. Height 35cm (14in) / H4-9

PLANTING ADVICE
Don't cover the crowns
of heucheras or they can
rot away, and do divide
them regularly. They do
well in containers but are
more prone to vine weevil
attack when grown in 'soft'
compost.

from becoming exposed at ground level. These can make a
gourmet dinner for vine weevils. Divide in spring and pull away,
or cut, pieces of fleshy stem from the outer edges of the clump.
Trim them underneath emerging leaves and pot them up in a
50 per cent mixture of perlite and potting compost. Use coarse
grit if you prefer.

Heucheras come from North America and they divide
into woodlanders found in the forests of Eastern America and
mountainous species from the Cordilleran mountains close to
the west coast. Garden forms tend to be aligned to *H. americana*,
a woodland species with good foliage but insignificant flowers.
Recent hybridisation has concentrated on foliage, rather than
flower, and Terra Nova's Dan Heims has raised an amazing range
of foliage colours including bronzes, blacks, limes and oranges.
Height 25cm (10in) / H4-9

'MOONSHINE'

ACHILLEA

Silver filigree foliage underpinning a long succession of sharp-yellow domes.

'Moonshine', raised by Norfolk nurseryman Alan Bloom (1906–2005) in 1954, was a breakthrough achillea because it was the first one to repeat-flower in flushes from late spring until autumn. It had excellent foliage, finely divided and silvered, and it was clump-forming too. Most achilleas have dull green, feathery foliage and one flush of flower in the second half of summer. They also stray and all three traits are due to their yarrow heritage derived from *Achillea millefolium*. 'Moonshine' broke the mould, becoming a worldwide success, and it was one of Alan Bloom's two favourites, the other being a crocosmia called 'Lucifer'.

FOUR MORE PALLID-YELLOWS TO TRY

ANTHEMIS TINCTORIA 'E.C. BUXTON'

At first, this lucid yellow daisy sulks and droops its petals, but by midday the flowers look like spinning plates. Found in this gentleman's rain-lashed North Wales garden in the 1920s and impervious to damp conditions. Height 60–90cm (2–3ft) / H3–7

HELIANTHUS 'LEMON QUEEN'

This tall lemon-yellow daisy fails in dry summers. However, if it gets enough summer moisture, it's truly terrific in August. The flowers follow the sun, so place carefully. Height 1.8m (6ft) / H3–9

HEMEROCALLIS 'WHICHFORD'

This British-bred fragrant daylily, raised by Harry Randall in 1960, produces clusters of pale-yellow flowers in June. The neat foliage is darkly shaded at the base, and it dies away in winter. Height 80cm (32in) / H3–9

HELIANTHEMUM 'WISLEY PRIMROSE'

An edging plant for sun, with grey-green foliage and plenty of primrose-yellow saucers in summer. Height 25cm (10in) / H3–9

Bloom's vast plant knowledge and plant collection taught him where the gaps were in the horticultural trade and he set about plugging them. In the years preceding the Second World War, he visited botanic gardens and German and Dutch nurseries. On one such trip he collected a straggly yellow hybrid achillea, with *A. taygetea* blood. It died in the first winter. However, Alan had collected seeds – although almost all of the seedlings died too. 'Moonshine' was one of a few survivors. Alan used these straggly achillea seedlings to plug a gap in one of his island beds in his Dell Garden at Bressingham. He told me that they could have easily ended up on the compost heap.

'Moonshine' is still a star plant seventy years later, due to its free-flowering habit. The pallid yellow domes are held on woody stems and, if you plant three or five, they spin through a sunny border like a golden thread.

Height 60cm (2ft) / H3–8

PLANTING ADVICE

'Moonshine' was a perfect nurseryman's plant, because it could be propagated from pieces pulled away from the main plant. Doing this, along with deadheading, encourages more flowers and lengthens the lifespan of the plant by preventing woodiness. Give it a bright position, reasonable drainage and avoid any rich feeding.

'SUMMER NIGHTS'

HELIOPSIS HELIANTHOIDES VAR. SCABRA

Red-cupped yellow daisies held on dark, branching stems in midsummer.

This is the first warm-yellow daisy of the year and announces that summer is wearing on. Used in autumnal borders, it kickstarts the display and the red centres of each daisy, along with the dark stems and foliage, smoulder like hot lava. It will continue to fizzle until September and by then it will be mingling with early asters, foamy grasses and a multitude of later yellow daisies.

Division is not easily achieved, in my experience, so most commercially available plants are grown from seed-raised plugs. This plant hates wet feet. But don't we all!

The species is a native of eastern and central North America, from east Saskatchewan to Newfoundland and as far south as Texas, New Mexico and Georgia. Known as the smooth ox-eye, it's definitely a prairie plant, but it's also drought-tolerant. The cone-shaped centre disc is hummingbird- or bee-pollinated in the wild, so it's worth deadheading this in the garden because it will flower for far longer. Position it where you can reach it, and weave several through your planting to give little pops of warm gold and red colour. There are double forms, like 'Summer Sun', and semi-doubles too. 'Summer Nights' was a selection made by Dale Hendricks of North Creek Nurseries, USA. It was introduced by Jelitto Perennial Seeds in 2004 and they've gone on to select golden-red 'Burning Hearts' and the orange-red 'Bleeding Hearts'. Often grown as a cut flower.
Height 1m (39in) / H4–9

FOUR MORE YELLOW DAISIES TO TRY

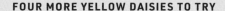

HELIANTHUS 'CAPENOCH STAR'

This waist-high daisy was a 1938 introduction, named after an estate in Dumfries, Scotland. The neat ring of bright-yellow ray petals surrounds a fuzzy cone of curly petaloid stamens and it flowers for two months. Height 1.2m (4ft) / H4–9

RUDBECKIA TRILOBA

A short-lived perennial that comes into its own from September onwards, when the rounded bush is smothered in small yellow flowers, each with a rich-brown middle.
Height 1m (39in) / H4–8

RUDBECKIA FULGIDA VAR. DEAMII 'DEAM'S CONEFLOWER'

The best clear-yellow Black-eyed Susan of all, with crisp 'oh-so-neat' brown-centred daisies from August until late autumn. The foliage never goes shabby. Height 60cm (2ft) / H3–10

RUDBECKIA LACINIATA 'HERBSTSONNE'

Columnar, slender daisy with large green-middled yellow daisies held way above the clump of foliage. Height 1.5m (5ft) / H3–10

AMSONIA TABERNAEMONTANA

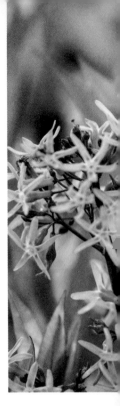

A May gap-filler, with clusters of pale slate-blue stars held atop dark stems.

American gardeners love their Eastern blue stars, a native flower that's endemic to much of the United States, according to horticultural guru Allan M. Armitage. However, amsonias are very underused in European gardens, despite offering something for every season. The late-spring spears rise from the ground like charred asparagus, and in no time at all the willow-like foliage appears, each leaf neatly defined by a strong-white midrib. Think Audrey Hepburn chic. The stems

FOUR MORE AMSONIAS TO TRY

AMSONIA TABERNAEMONTANA 'BLUE ICE'

A shorter, roughly 60cm (2ft) amsonia with brighter blue flowers. Not enough billow and flounce for me! Height 60cm (24in) / H4-9

AMSONIA TABERNAEMONTANA 'STORM CLOUD'

A compact, sultry selection with darker stems and brighter blue flowers. It was awarded the 2024 Perennial Plant of the Year by Proven Winners and is readily available. Height up to 90cm (3ft) / H4-9

AMSONIA HUBRICHTII

Narrow, ferny foliage and pale grey-blue mounds of flower, with a floppier habit than the others. Height up to 90cm (3ft) / H4-9

AMSONIA 'ERNST PAGELS'

A fine hybrid between *A. ciliata* and *A. hubrichtii*, named after the Dutch nurseryman Ernst Pagels. This narrow-leafed amsonia splays out more rigid stems topped by tight lilac-blue clusters of stars. Colours up in autumn. Height 90cm (3ft) / H4-9

PLANTING ADVICE
Amsonias are slow-fuse plants: they take their time to form a clump. Planting three in a group speeds the process up. Rich, moisture-retentive soil also helps, although once established amsonias have no trouble tolerating summer drought and humidity. They are rarely troubled by slugs or deer, perhaps due their milky sap. They all like an open site. Cut back in late autumn.

stay soot-black and then the clusters of buds open and produce a multitude of almost grey-blue, starry flowers. In late summer, narrow seedpods form and then darken as autumn approaches, just as the foliage turns butter-yellow.

I think amsonias need to flow and billow, but stay upright as well, and I like their flowers to be the colour of a dull summer sky on a humid day, as depicted by J.M.W. Turner. Then they'll add extra softness to May, a season of verdant foliage. *Amsonia tabernaemontana*, the Eastern blue star, has green foliage with prominent white midribs: imagine Fred Astaire dressed in spats and patent pumps dancing with Audrey. The pyramidal clusters of soft grey-blue flowers, so good in evening light, last for many a week from mid-May onwards, a time when little else is out. Bright shade, with afternoon sun, is a good position because it won't dilute the gloaming colour to a dirty white.
Height 90cm (3ft) / H4–9

'SAHIN'S EARLY FLOWERER'

HELENIUM

Dazzling orange and yellow ray petals set around a brown velvet dome.

Heleniums, or sneezeweeds, are a challenge for many gardeners because the most commonly grown forms are allied to *Helenium autumnale*, a species that needs moist soil throughout the growing season. Moist summer soil can no longer be guaranteed and many a gardener, including me, has chosen to avoid them. 'Sahin's Early Flowerer', a sterile hybrid, thrives far better. It has more vigour and tenacity and it repeat-flowers from May until September, because visiting bees are unable to pollinate it successfully. The dazzling streaked yellow and orange flowers have variable colour combinations.

PLANTING ADVICE

This dazzles in a border, so you'll need to plant more than one to get the full effect of the 'jumping-gene' flowers, which come in a range of patterns and streaks. Plant in spring, if possible. Deadheading will promote more flowers faster and it looks neater, but this amazing helenium will still produce more flowers even if neglected. Don't over-fertilise. Cut back in autumn, divide in spring – but only if you wish to.

This plant was a chance discovery by the late Dutch seedsman, Kaas Sahin. He found it growing in a field full of thousands of seed-raised heleniums. It stood out because it flowered precociously early and carried on doing so until late. Kaas recognised its potential but had little time to exploit it, so he passed it on to Bob Brown of Cotswold Garden Flowers. Bob named it then micro-propagated it and it's one of the best perennials to emerge within the last twenty-five years.

Although some gardeners find orange difficult to place, it's the touchpaper colour for blues and purples because these colours are on opposite sides of the colour wheel. 'Sahin's Early Flowerer' enhances summer-flowering blue salvias, agastaches and nepetas and it's one of the first vibrant daisies to worship the summer sun. Be bold with orange.
Height 90cm (3ft) / H3–9

FIVE MORE HELENIUMS TO TRY

'MOERHEIM BEAUTY'
Bred in the 1930s by Ruys, but still unbeatable, with a double layer of mahogany-red petals. These flounce downwards like a flamenco dancer's skirt in mid-flourish. Height 90cm (3ft) / H3–9

'FEUERSIEGEL'
A fiery lightning bolt, with deep-yellow flowers marked in blazing red. Means 'fire seal'. Height 1.2–1.8m (4–6ft) / H3–9

'FLAMMENSPIEL'
Meaning 'dancing flames', with orange-red streaked with yellow. Starts in the last few days of August. Height up to 1.2m (4ft) / H3–9

'WALTRAUT'
Shaggy orange flowers earlier in the season. Means 'strong'. Height 90cm (3ft) / H3–9

'GARTENSONNE'
Tall, with primrose-yellow petals set around a brown middle that slowly develops yellow fuzzy tips after bees visit. Meaning 'garden sun'. Height up to 1.5m (5ft) / H3–9

'ROMA'

ASTRANTIA

A candy-pink Hattie's pincushion that offers flower power over many months.

'Roma', named by the Dutch landscape architect Piet Oudolf, was a seedling from *A. major* 'Ruby Wedding' and that's the main problem with most forms of *Astrantia minor*. They produce nuisance seedlings and then stop flowering. The resulting progeny are usually inferior to the parent. The darker reds and rich pinks have a flourish in May and then they may throw a few flowers in autumn. However, 'Roma' is sterile, so there are no unwanted seedlings, and the flowers remain on the plant for far longer and keep on arriving. As a result, 'Roma' can be in flower for many months.

THREE MORE ASTRANTIAS TO TRY

ASTRANTIA 'BUCKLAND'

Found by Keith and the late Ros Wiley at The Garden House in Buckland Monachorum and thought to be a hybrid between *A. maxima* and *A. major*. Small pale-pink flowers aplenty. No seedlings. Height 50cm (20in) / H4–9

ASTRANTIA MAXIMA – THE HELLEBORE-LEAVED MASTERWORT

Maxima's the clue here, because the mid-pink flowers are larger than most astrantias with the exception of 'Shaggy'. It's not easy, because this lush meadow plant demands rich, moist soil. Blooms once between May and July. Has running roots – but it isn't a nuisance. Height 80cm (32in) / H4–9

ASTRANTIA MAJOR SUBSP. INVOLUCRATA 'SHAGGY'

Named by Margery Fish, who gardened partly on rich Somerset soil and partly on dry poor soil. It was grown in the cottage gardens in Gloucestershire and the large white flowers have green tips. Needs lavish conditions to produce that large flower. Height 70cm (28in) / H4–9

Astrantias are members of the cow parsley family, the Apiaceae, but their radiating stems are concertinaed and encased in a neat circlet of jagged bracts. Their umbels of tiny flowers are suited to small-mouthed pollinators, such as hoverflies, and these are useful garden allies, because their voracious larvae eat pests such as aphids. Many of the most desirable astrantias are dark red or claret-coloured and they are exquisite. However, 'Roma' stands out better, due to its almost candy-pink colouring. Height 50cm (20in) / H4–9

PLANTING ADVICE
Astrantias make woody root stocks, in time, and this can create bare patches. Regular division in spring every three years is advisable to prevent this happening. Cut back the faded flower stems religiously too, because they brown horribly as they fade. Most dark astrantias prefer rich, moisture-retentive soil and they will not re-bloom again in autumn on drier soil. However, 'Roma' does not seem to mind and neither does 'Buckland', another paler-pink sterile. They both do well, full stop, in sun or bright shade.

'TOTALLY TANGERINE'

GEUM

Perfection and poise, along with a long succession of apricot-orange tissue-paper flowers held on airy stems.

If I was allowed to take one plant to a desert island, this is definitely the one I would choose, because it hints at summer before it's really arrived. It was the Chelsea Plant of the Year of 2010 and it has appeared there ever since in countless show gardens purely because it performs so well in May. It's sterile, so the tall wands of soft-orange flowers appear in flushes until late in the year, making it the longest flowering geum of all. It was deliberately raised by Tim Crowther of Walberton Nursery over a period of years. He meticulously crossed and back-crossed *Geum rivale*, the female woodland parent, with the sun-loving *Geum chiloense* 'Mrs J. Bradshaw'. 'Totally Tangerine' was named in 1999, although it was originally listed as 'Tim's Tangerine'.

THREE MORE GEUMS TO TRY

'LEMON DROPS'

A woodland shade-lover found in Beth Chatto's Essex garden, with drooping heads of greenish-yellow buds that open to yellow. Deadhead after flowering and place near blues such as *Pulmonaria* 'Blue Ensign'. Height 30cm (12in) / H5–7

GEUM 'BELL BANK'

A woodland shade-lover discovered by the late television gardener Geoffrey Smith (1928–2009) and named after a damp area of his garden at Kettlesing in Yorkshire. Dark buds open to reveal very large copper-pink semi-double flowers in April and May. At first these hang their heads demurely, as do many early flowering geums, but then they straighten up and open wider before fading to paler pink. Height 60cm (2ft) / H5–9

'HILLTOP BEACON'

This pinkish-orange geum produces double flowers for twelve weeks from mid-April onwards. Height 80cm (32in) / H5–9

Regular division keeps
geums alive and helps
prevent ground weevil
attacking the old, exposed
stems. 'Totally Tangerine'
will need dividing every
three or four years, in
spring. Potting up the
pieces and then replanting
works best. Dry summers
do affect its ability to
flower again.

The spent seedheads resemble ginger spiders in autumn
sunlight, so it's worth leaving some of the flowering stems in
place. Once autumnal rains come, it will re-bloom again but never
as generously as it did so in May. There are fifty species of *Geum*
found in a wide area, principally North and South America, Asia,
New Zealand and Africa. They are very diverse. Spring-flowering
G. rivale cultivars have shy, nodding flowers in gentle colours.
Upward-facing oranges and clear yellows appear from May
onwards, when much of the garden is full of pastel blues, lemons
and pinks. These can be hard to place, but the soft tangerine
flowers of 'Totally Tangerine' will slot in anywhere.
Height 90cm (3ft) / H5-7

'PURRSIAN BLUE'

NEPETA X FAASSENII

A tidy, compact catmint with masses of dainty blue flowers held above small grey-green leaves.

Aromatic catmints thrive in sunny positions in well-drained soil and they don't need feeding. It makes them floppy. They flower for many months and the bees adore them. The bushier ones can be cut back to nothing in late July, when they begin to look ragged. They send out new shoots within days and this harsh treatment makes them flower later into the year, so you get months of colour. Shearing them back also keeps them vigorous, because there's no opportunity to get woody.

FOUR MORE CATMINTS TO TRY

NEPETA RACEMOSA 'WALKER'S LOW'

This is the natural successor to 'Six Hill's Giant' and, although labelled low, it's not short. It was introduced in the 1980s by Walker's Low Nursery and named Perennial of the Year in 2007 by the US Perennial Plant Association. The lavender-blue flowers just keep on coming. Height 75cm (2.5ft) / H5–9

NEPETA 'HILL GROUNDS'

Found as a seedling in the late Janet Cropley's Northamptonshire garden of the same name, this plant has a gently sprawling habit. The exceptional foliage is grey-green and the larger than average flowers are violet-blue. Height 30cm (12in) / H5–9

NEPETA 'WEINHEIM SUMMER BLUES'

This green-leaved catmint has whorls of large purple-blue flowers with lightly spotted lips. It does run but 'Weinheim Big Blue' is said to be even more vigorous, according to the RHS trial panel. Height 80cm (32in) / H5–9

NEPETA 'BLUE DRAGON'

An introduction from Janet Egger at Terra Nova Nurseries, this upright form has notably large, blue-violet flowers and bright-green foliage that turns a golden red in autumn. Copes well in damper areas and flowers a little later than most. Height 60cm (2ft) / H5–9

The go-to catmint used to be 'Six Hills Giant', but this tends to sprawl and flop. Many plants sold are often wrongly labelled, as the RHS Wisley Nepeta Trial (2018–2021) proved. 'Purrsian Blue', one of the best on the trial, is compact and the mauve-blue flowers are held in purple-blue calices above small grey-green leaves. It can be grown in containers, or used to edge a sunny border, or it can be woven through to unite a planting scheme. It was introduced by Walters Gardens in the US state of Michigan in 2013 and became an international success. It's very compact, although it doubles in size in its second year, but still keeps its neat, rounded habit.
Height 45cm (18in) / H5–9

PLANTING ADVICE
Nepetas are drought-tolerant and deer-resistant and they flower for many months, pleasing lots of pollinators. Plant in spring, if possible, in a sunny, warm position. Divide in spring, if you need to. If they get battered by hard winter weather, cut them back when spring proper arrives.

'DUCHESSE DE NEMOURS'

PAEONIA LACTIFLORA

Lemon-scented, fully double cream flowers with a slight touch of pale yellow – catching the softness of an early summer's day.
This is one of several pastel-coloured, fragrant peonies raised from *Paeonia lactiflora*, a Chinese species with glossy foliage and several buds per stem. They have far more flower power than the cottage garden peony (*P. officinalis*), which only puts out one flower per stem for only one week in May. Many of these scented 'lactiflora' pastel peonies were bred in France during the mid-19th century, principally for the cut flower trade. 'Duchesse de Nemours' was bred by Calot in 1856 and many have names that could be straight out of a Parisian telephone directory.

Paler peonies are sweetly scented, so French breeders concentrated on producing cream, white and pink varieties because their customers wanted fragrance in the vase as well as the garden. They became staple cutting-garden plants, so they

FOUR MORE PEONIES TO TRY

PAEONIA 'BUCKEYE BELLE'
An early free-flowering double red with dark foliage. Needs warmth! Height 75cm (2.5ft) / H3-8

PAEONIA 'CORAL SUNSET'
This is a chameleon. Apricot cupped flowers, with up to fifty petals, open but quickly fade to shades of clotted cream providing a peaches and cream confection. Height 90cm (3ft) / H3-8

PAEONIA 'BARTZELLA'
An intersectional, or Itoh peony, with woody stems inherited from its tree peony genes and masses of large lemon-yellow flowers. Known as Godzilla in my garden – for its brash strength. Height 90cm (3ft) / H4-8

PAEONIA LACTIFLORA 'FESTIVA MAXIMA'
Mid- to late-season with double flowers that fade to cream, with an occasional vivid, maroon-red blaze. Height 1m (39in) / H3-8

were planted across the world for commercial reasons. They were picked when the flowers were the size of marshmallows. Breeding moved on from France when Kelways of England began selecting more compact varieties in the first part of the 20th century. James Kelway collected *P. mascula* subsp. *mascula* from a monastic garden on Steep Holm, a small island off his native Somerset. These shorter peonies could be shipped around the world.

American breeders went in another direction. They hybridised colourful species to produce May-flowering cultivars in a wider range of colours, including reds and corals. However, these lack fragrance and they often need a warm site. Peonies mix well with roses. They both share coppery young foliage and both prefer good soil and a bright position. Most peonies flower just before the roses. Should they overlap, though, the flowers complement each other.
Height 90cm (3ft) / H3–8

PLANTING ADVICE
Peonies are extremely hardy and long-lived if they're given a bright position in good soil. Those with sumptuous flowers need staking, but apart from that peonies are trouble free. Most herbaceous peonies enter dormancy in August, looking shabby, so don't plant them next to late summer and early autumn flowering performers. Cut them back by September and divide in October. The woody roots hate being waterlogged.

'OSTFRIESLAND'

SALVIA NEMOROSA

Flickering gas jets of upright blue and red tapers, flattered by soft sage-green foliage.

This robust and long-lived hardy salvia is something to write home about, because so many so-called hardy salvias fade away over winter due to wet conditions. This one definitely doesn't. The strong blue, bee-friendly flowers have a touch of red, adding vibrance to June and July, and in dry summers the stems turn purple too. Raised by Dutchman Ernst Pagels in 1949, and named after his homeland, it's been widely used by Piet Oudolf in countless prairie planting schemes.

The young Oudolf visited Ernst Pagels' Nursery at Leer, because it was close to the German–Dutch border. Pagels, a man who bred stiff-stemmed perennials and grasses that were able to

FOUR MORE PAGEL'S PLANTS TO TRY

BETONICA OFFICINALIS 'HUMMELO' (PREVIOUSLY STACHYS)

Raised by Pagels and named in honour of his friend Piet Oudolf's home, this purple-pink betony produces a mass of flower spikes in early summer. Height 90cm (3ft) / H4–9

HYLOTELEPHIUM TELEPHIUM (ATROPURPUREUM GROUP) 'KARFUNKELSTEIN'

One of the best mound-forming border sedums for smoky foliage and good pink-red flower. Similar to 'Xenox', with dusky foliage from early on. Height up to 45cm (18in) / H4–9

ACHILLEA 'WALTHER FUNCKE'

A hybrid between *A. millefolium* and *A. filipendula*. This stiff-stemmed achillea has tiny pin pricks of yellow in the orange-red flowers, which appear between June and September. Height 60cm (2ft) / H3–8

VERONICASTRUM VIRGINICUM 'LAVENDELTURM'

One of the best tall verticals for late summer, with upright long spires of pinkish-lavender flowers held above whorls of green leaf. Beloved by bees. Height up to 1.5m (5ft), perhaps more on good soil / H3–9

PLANTING ADVICE
Like every other aromatic
plant with pungent foliage,
'Ostfriesland' needs a sunny
position. Cut it back in
the spring, not in autumn,
because the top growth
protects the plant's roots.
It's best divided in spring,
just as regrowth occurs at
the base, but only if you
need to. This bushy sage
stays compact so it can
be left for many a year. It
prefers poor soil, like other
pungent plants, so no rich
feeds please.

fade into winter beautifully, was in turn mentored by Karl Foerster
of Potsdam in Germany. These three horticulturalists, one Dutch
and two German, changed the way we look at our gardens more
than anyone else. Their aim was *creating* winter interest, good
flowers and stiff stems that needed no staking.

Pagels selected for cold-hardiness, a long-flowering period,
good seedheads and disease resistance. Ease of propagation was
another consideration. *Salvia* 'Ostfriesland' is one of his finest
plants and it was selected from seeds given to him by Karl Foerster
in 1949, the year Pagels founded his Leer nursery. 'Blauhügel',
meaning 'Blue Hill', and 'Tänzerin', meaning 'Dancer', followed on.
Height 50cm (20in) / H5–9

SPANISH OAT GRASS

STIPA GIGANTEA

A wizard plant, straight out of the box of exploding fireworks, providing a tall, transparent metallic veil.

Like all stipas, this grass needs a sunny position and reasonable drainage. It makes a real impact by early summer, unfurling golden quills just as the purple alliums open in May. By midsummer's day *S. gigantea* is in full flight and it will continue to dazzle until mid-autumn, offering months of interest. The evergreen tussock of fine silvery foliage, which rarely gets over-battered by wintry weather, gives it an early advantage. Trimming the sides of the foliage in autumn keeps the sides of the tussock tight, although the top growth is left to overwinter. Think top hat!

Always plant in spring and, whatever you do, never divide your stipas. They really resent it and they will probably turn up their toes in shock. Find a key position in bright sun, preferably a place that you look through. Remove the heads when they break up, perhaps in late autumn, and then clip the foliage into that top hat shape mentioned before.

Stipa gigantea is variable, in both height and form. Seek out Ernst Pagels' 'Gold Fontaene' for its taller, more lavish, golden heads. It shimmers over many months, cascading over perennials and adding sparkle to summer-flowering purples and dark blues. The golden fountain also picks up the yellow and orange crocosmias and it's still working its magic as the autumn daisies open. Use it as a backbone in the midst of a large border, or run it through the border to pull the eye along. If room's tight, plant one as a specimen, but it deserves to be planted in numbers!

I love the combination of gun-metal grey foliage topped by a golden sheath, but I am not keen on the compact forms like 'Pixie' and 'Goldilocks' because they have all-green foliage.
Height up to 2m (6.5ft) / H5–10

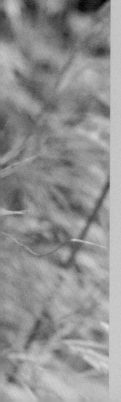

FIVE MORE GRASSES OF NOTE TO TRY

JARAVA ICHU – SYN. PERUVIAN FEATHER GRASS
Grown for its tapering straw-white feathery heads, which move and sway like bleached-blond hair. Height 60cm (2ft) – taller in good soil / H8–9

MELICA UNIFLORA F. ALBIDA
A shade-loving melick producing bright green leaves followed by a sprinkling of rice grains by early summer. Fab with ferns! Self-seeds, but I can never get enough of this spring-fresh grass, so I don't care. Height 30cm (12in) / H7–9

CORTADERIA RICHARDII
Tall and feathery with graceful golden heads that rise to great heights, shimmying as they go. Seedlings will bother those in warmer gardens. Height up to 3m (10ft) / H7–8

PENNISETUM MACROURUM FOXTAIL GRASS
A seed-raised perennial, with green leaves topped by slim, cylindrical buff-white to beige heads. Flowers in autumn in cool gardens. Height 1.2m (4ft) / H8–10

MOLINIA CAERULEA SUBSP. *ARUNDINACEA* 'TRANSPARENT'
Tall and sheath-like, with airy heads of dark beads held on tall stems, from late summer. Use it as a veil: don't be afraid to give it a front-of-border position. Cut back before the year ends, because the stems crash. Height 1.5m (5ft) / H5–9

JAPANESE HAKONE GRASS

HAKONECHLOA MACRA

Curtseying foliage that flounces down to the ground, topped by feathery, dark heads in early autumn.

These deciduous grasses come in various foliage colours, from green and yellow, to all-gold, to pure green. They're almost deciduous, because the foliage turns to brown paper by late winter. Cut back then and new foliage appears in late spring. By May they're cascading, as if they're being poured out from an overfull pot by the sorcerer's apprentice. The linear leaves seem to arrange themselves gracefully and stay that way, so they can spill over an edge, or fountain out of a container, or pop

FIVE MORE HAKONE GRASSES TO TRY

'AUREOLA'
Gold- and green-striped foliage, forming dense hummocks, that change to chartreuse-green, according to grass expert Neil Lucas of Knoll Gardens. Height up to 40cm (16in) / H4–8

'ALBOVARIEGATA'
Bright-white and cream-striped foliage that lights up dull corners of the garden. Less likely to scorch than 'All Gold'. Height up to 40cm (16in) / H4–8

'NICOLAS'
Compact Hakone grass with all-green leaves that colour up to provide touches of wine-red in autumn. Lovely and fresh in spring. Height 30cm (12in) / H4–8

'SUNFLARE'
This one defies logic, because the completely chartreuse-green leaves only develop yellow highlights in brighter positions. Height up to 40cm (16in) / H4–8

'MULLED WINE'
A new Hakone grass, compact with golden- and green-striped leaves. Named because the new foliage colours to burgundy red during the season, showing all three colours at once. Height up to 40cm (16in) / H4–8

up towards the front of the border. They billow in the breeze, but rather like well-lacquered hair they manage to remain immaculate. These grasses spread slowly, getting denser rather than wider, adding texture to shadier, cool areas of the garden.

They often thrive under trees and shrubs and this benign position suits them well, because this grass dislikes winter wet as so many Asian plants do. They can be held back by extreme cold. In the wild this grass is found naturally in moist, mountainous, wooded areas of central Japan, thriving around Mount Hakone. Hence its name, Hakone Grass. The vivid golden forms light up dark corners of the garden really well.

Height up to 40cm (16in) / H4–8

PLANTING ADVICE
Use in shade or bright shade, not full sun. The more gold on the leaf, the more the leaves generally scorch in summer sun. The foliage breaks down in midwinter and that's the time to cut it back, before it covers the garden in fragments of papery-brown leaf. Watch out for all-green reversion in the variegated forms. Plant and divide in spring. They're really hardy.

'PRICHARD'S VARIETY'

CAMPANULA LACTIFLORA

A handsome mophead of true-blue flowers that can flop a little.

Campanulas, or bellflowers, are a varied bunch ranging from small alpine right up to the metre high 'Prichard's Variety'. Many are not long-lived, because they seem to flower themselves to death. Others roam off in search of better soil. This is one of the very best, staying put and flowering well. The true form has been around

PLANTING ADVICE
This campanula needs good soil that's reasonably moist. After time the root stock gets woody and gnarled and division at this stage is vital. It can lead to losses, so the best way to propagate is from cuttings taken in April.

since Maurice Prichard's Riverslea Nursery in Chichester, West Sussex, released it in back in the 1920s. However, there are some imposters on sale, taller and floppier with paler blue flowers. The true 'Prichard's Variety' should have Cadbury-blue flowers.

Campanula lactiflora, the milk-flowered bellflower, is native to areas around the Black Sea. The first flowers appear at the top, but once these are deadheaded more are produced in the leaf axils, so this plant can give weeks of flower. You can also Chelsea-chop the stems, in the third week of May, to delay flowering time. Then the flowers follow the main flush of roses. The vivid blue flowers and long-flowering season, along with its ability to be manipulated by gardening tweaks, make it an excellent addition to summer-flowering borders. And it doesn't need staking. Height 60–90cm (2–3ft) / H4–8

FIVE MORE CAMPANULAS TO TRY

CAMPANULA 'KENT BELLE'

Bred by Elizabeth Strangman in Kent, hence the name, this *C. takesimana x C. latifolia hybrid* has heart-shaped leaves and deep-violet tubular bells, plus a running root stock. Height 75cm (2.5ft) / H5–9

CAMPANULA LACTIFLORA 'LODDON ANNA'

A shorter pale-pink version of 'Prichard's Variety', from Tommy Carlile's Loddon Nursery in Twyford, Berkshire. Dates from the 1920s, and it doesn't seem to develop a woody root stock. Height up to 90cm (3ft) / H4–8

CAMPANULA GLOMERATA 'CAROLINE'

A mauve-pink short-lived campanula, which favours limey soil and dappled shade.

Beguiling, but less vigorous than the blue forms. Height up to 60cm (2ft) / H3–8

CAMPANULA PERSICIFOLIA 'PRIDE OF EXMOUTH'

A double-flowered heritage form of the peach-leaved bellflower, found in this south Devon town. The powder-blue frilly flowers, one cup inside another, have a cottage-garden charm. The creamy stamen-rich centre looks almost edible. Height 90cm (3ft) / H3–8

CAMPANULA CARPATICA 'BLAUE CLIPS'

The German selection of this Carpathian Mountains species is more alpine in habit, but the mid-violet flowers appear over a long period. And it's long-lived. Height 20cm (8in) H4–9

'SPARK'S VARIETY'

ACONITUM

Small pops of midnight-blue flowers – all summer long.

This aconitum deserves a place in every garden, because it produces its own radio masts! The main stem rises to waist-height, up to 1.2m (4ft), producing side shoots that leave the main stem at a 45-degree angle. It is the only aconitum, or monkshood, to have this growth habit. The side shoots develop buds at different times, so there's never a solid spire, just a fairy-light explosion. It prefers good soil, but it's strong enough to push up through an herbaceous border year after year. Better still, this midnight-blue aconitum has good foliage throughout the growing season. The pinks and whites are not nearly as good in the garden.

Aconitums' flowers attract the bees, so most produce copious numbers of seeds, although I believe 'Spark's Variety' is sterile in my garden. You need to remove the hooded seedheads at the green stage, because seedlings are invariably inferior

FOUR MORE ACONITUMS TO TRY

ACONITUM CARMICHAELII VAR. TRUPPELIANUM 'KELMSCOTT'

A little later and a little less vibrant, but still a cracker in autumnal light. Height 1.2m (4ft) / H4–8

ACONITUM X CAMMARUM 'BICOLOR'

An earlier monkshood, with white flowers colour-washed in violet-blue usually after midsummer. Height 1.2m (4ft) / H4–8

ACONITUM X BICOLOR 'BRESSINGHAM SPIRE'

Raised by Alan Bloom in 1957, this violet-blue taper flowers in the second half of summer. Highly divided 'napellus' type foliage. Height 90cm (3ft) / H4–8

ACONITUM 'STAINLESS STEEL'

A summer-flowering, silver-blue to grey aconitum that shines in evening light. Looks best in dappled light. Height 1.2m (4ft) / H4–8

PLANTING ADVICE
Aconitums are long-lived
plants that come up and
flower well every year, so
they don't need regular
division. If you do divide,
do so in late winter or
early spring. Wear gloves
and wash your hands
well. These stately plants
tolerate winter wet well
and some shade. Slugs
avoid them.

to the parent. The later autumn-flowering monkshoods are
invaluable with late-season grasses such as *Miscanthus sinensis*
'Silberfeder'.

'Spark's' was popularised by Maurice Prichard of the
Riverslea Nursery in 1898, so it's endured for over a century.
I'm not sure whether it was named after a person, or named for
its ability to create floral sparks. All aconitums are toxic and
this goes for the roots, stems and flowers. Its common name,
wolfbane, indicates that it had enough poison to kill a wolf.
Always wear gloves when handling this plant, whether it's
division or cutting back, and wash your hands afterwards. Some
people are more susceptible than others. For all that, aconitums
are really worth growing for their blue flowers.
Height 1.4m (4.5ft) / H4–8

'MÖNCH'

ASTER X FRIKARTII

A cloud of large lavender-blue daises, that are freeze-framed from July until mid-autumn, on this superstar.

This drought-tolerant aster is in my top five perennials because it performs from July until October, whatever the weather. It's a 1918 hybrid produced by a Swiss breeder named Karl Ludwig Frikart, using two European species. One was the blue-flowered Italian aster or *Aster amellus*, native to central and southeastern Europe. The other was a pink-flowered, short Himalayan species named Thompson's aster or *Aster thomsonii*. The latter is now thought extinct.

PLANTING ADVICE
Best planted in spring, if possible, and this is also the time to divide – but only if you need more. This division and planting advice goes for any perennial that flowers after midsummer. Spring is best!

Frikart's hybrids became a great success because they had hybrid vigour and they were sterile, so they flowered for many weeks. They were drought-tolerant and yet they came through winters and they continue to do this, even though our winters are often very wet. Three of his eleven are still with us today, the widely available 'Mönch', and the less so 'Eiger' and 'Wunder von Stäfa'. They're all named after mountains in the Bernese Oberland.

The stems are not ramrod-straight and this slightly loose habit makes them perfect front-of-border plants because they do a low curtsey over hard edges. Their slightly hairy green foliage is just as enduring as the bee- and butterfly-friendly flowers.
Height up to 90cm (3ft) / H4–9

FIVE MORE MILDEW-FREE ASTERS TO TRY

ASTER AMELLUS 'VEILCHENKÖNIGIN'

This Italian aster has long-lashed flowers in violet-blue, like Twiggy's mascaraed eyelashes, on stiffer, slightly shorter stems. These Italian asters pre-empt and overlap with 'Mönch'. Height 60cm (2ft) / H4–7

ASTER AMELLUS 'KING GEORGE'

A softer mauve, rather than purple, showing off those small orange-yellow middles. Height 60cm (2ft) / H4–7

ASTER X FRIKARTII 'WUNDER VON STÄFA'

Very similar to 'Mönch', with lavender-blue daisies. Another good performer. Height 90cm (3ft) / H4–9

ASTER AMELLUS 'ROSA ERFÜLLUNG'

Raised by Karl Foerster, a more erect aster with deep-pink flowers and much smaller leaves. Height 60cm (2ft) / H4–7

ASTER AMELLUS 'SONORA'

A recent, purple Italian aster with a central yellow disc that reddens once the bees have visited. Greyer foliage. Height 60cm (2ft) / H4–7

'GEORGE DAVISON'

CROCOSMIA X CROCOSMILIFLORA

Branching heads of early, warm-yellow flowers – vigorous but not aggressive.

Crocosmias have a South African provenance and this means that they enjoy rainfall in the growing season. They will not flower well in dry summers, despite the fact that they are often listed as drought-tolerant. They share this characteristic with other South African plants including agapanthus, dierama and kniphofia. However, crocosmias respond to being watered and they have pigment-packed petals and sword-shaped foliage. They provide a contrast with the hummock shape most perennials have, so their sword-shaped foliage adds much to the border.

'George Davison' was named in 1900 and it was raised in Norfolk, which was the epicentre of crocosmia breeding in the

FIVE MORE CROCOSMIAS TO TRY

'STAR OF THE EAST'

A languid crocosmia, with enormous golden flowers. It seems to surprise you by popping up every now and then. Height 90cm (3ft) / H8–9

'LUCIFER'

July-flowering red crocosmia, with luscious green foliage, but it's an aggressive grower. Place it carefully. Height up to 1.5m (5ft) / H5–9

'BRIGHT EYES'

Green foliage and brown and orange flowers, like a miniature 'Emily Mackenzie'. Protect in winter. Height up to 60cm (2ft) / H7–9

'PAUL'S BEST YELLOW'

From Ken Ridgely, with warm-yellow outward-facing flowers that open wide. A striking crocosmia. Height 90cm (3ft) / H6–9

'LIMPOPO'

This orange-apricot, part of Paul Lewis's African Series, is a strong, hardy crocosmia, shorter than many. Height up to 90cm (3ft) / H6–9

PLANTING ADVICE
Crocosmias develop strings
of corms, with newest
corms being formed at
the top of the string every
year. They push themselves
down in the process and
in doing so become less
vulnerable to frost. Plant in
spring and cloche in their
first winter. Use the upper
corms when propagating.

early 20th century. George Davison, head gardener of Westwick
Hall in Norfolk, began to deliberately hybridise them, selecting
for flower size. Davison's new crocosmias were released by
Wallace and Company – the crème de la crème of nurseries then.
Wallace charged a high amount for these new introductions.
George Davison's finest is almost certainly the large flowered
burnt-orange 'Star of the East' launched in 1910.

However, 'Star of the East' isn't hardy for all. Worries about
hardiness in crocosmias in general deterred Alan Bloom of
Bressingham in Norfolk from working on them. The hard winter
of 1963, with frost penetrating to 30cm (12in) down, failed to see
them off. 'Lucifer' was one of his first seedlings and it's grown the
world over. Modern plant breeders select for hardiness and Paul
Lewis's Scorchio series, including 'Firestars', and David Tristam's
Warburton 'Bright Eyes' are recent additions.
Height 75cm (2.5ft) / H6–9

'GLOBEMASTER'

ALLIUM

Violet-mauve starry spheres that appear almost fluffy under the glare of the midday sun.

Taller alliums make a real statement in the garden. Their stems provide a strong vertical presence, linking ground to sky, and their substantial globular or domed flowerheads offer a contrast. They have excellent timing as well, flowering after the explosion of spring bulbs and woodlanders but before the main rose flush and the majority of summer perennials. They are all very bee-friendly and some produce copious amounts of 'nuisance' seeds. However, plant breeders have been raising bee-friendly, sterile hybrids. These stay in flower for longer and they don't produce viable seeds. These taller alliums do not need staking.

PLANTING ADVICE
Plant the bulbs in a bright position in September, to a depth of up to 20cm (8in) if possible. Use groups of three or five to create an impact. Hybrid alliums are expensive to produce, but they will endure for many years.

Many of these drumstick alliums perform in May, or very early June, and you need to hide the foliage with emerging perennials, because it tends to wither and fade just as the flowers appear. Globemaster, a 1971 hybrid between *A. cristophii* and *A. macleanii* bears huge violet globes that measure 25cm (10in) across in their first year. The starry florets overlap. As time progresses the heads get smaller. If you add a few new ones each year you'll get a natural mix of different sizes. Tulips also do this. Alliums are ornamental onions and they are very hardy, because many have a mountainside provenance. Deer, rabbits and rodents avoid them.

They can resist heavy rain and wind and many are purple or mauve, so they sparkle under the sun. Allium flowers cut well and their seedheads dry well too. They are equally at home in the vegetable garden and the flower garden and it's possible to get a succession of flowers using different hybrids.
Height 1m (39in) / H4–9

FIVE MORE ALLIUMS TO TRY

ALLIUM HOLLANDICUM 'PURPLE SENSATION'
The rich-purple flowers mix well with later tulips, but you must remove the flowerheads before the seeds dehisce.
Height 85cm (34in) / H4–7

'PURPLE RAIN'
This shorter, mauve-purple allium makes a good follow-up act for 'Purple Sensation'.
Height 85cm (34in) / H4–7

'AMBASSADOR'
An intense purple allium with 18cm-wide (7in) globes of starry florets for a five-week period. Height 1.2m (4ft) / H4–7

'GLADIATOR'
More of a pale rose-purple, this 1981 cross between *A. aflatunense* and *A. macleanii* bears 15cm-wide (6in) domes and each starry floret has a pinprick of spring green.
Height up to 1.2m (4ft) / H4–7

ALLIUM STIPITATUM 'MOUNT EVEREST'
This allium has cool-white stars centred in green and it persists for many weeks. Good in dappled shade, especially against a yew backdrop.
Height 90cm (3ft) / H4–7

'MONICA LYNDEN-BELL'

PHLOX PANICULATA

Clouds of pale-pink flowers to soften a summer's day.

Border phloxes are found naturally in open woodlands in various states of America, so most garden cultivars need at least six hours of sun and rich, organic soil. Named cultivars of *Phlox paniculata* are generally thirsty plants and most fail in summer droughts and develop mildew. 'Monica Lynden-Bell' is an exception. It was found in a chalky British garden in Hampshire in the 1970s and was popularised by British nurseryman Bob Brown of Cotswold Garden Flowers. Chalky soil is not normally suitable for garden phloxes, but this one (which probably arose as a seedling) tolerates drier conditions.

Monica is a mid-season phlox of medium height that reliably re-appears every year. Many garden phloxes need regular division to thrive and I find the blues and mauves the most

FOUR MORE BORDER PHLOXES TO TRY

P. PANICULATA 'BRIGHT EYES'
A very fragrant and resilient 1967 phlox with dark-pink centred, pale-pink flowers in the second half of summer. Height 90cm (3ft) / H4–9

P. PANICULATA 'DÜSTERLOHE'
Free-flowering purple with strong stems. Early and widely used in landscape and prairie planting. Karl Foerster, 1962. Height 90cm (3ft) / H4–9

P. PANICULATA 'NORAH LEIGH'
A tastefully variegated phlox with pale-pink flowers. Survives in all conditions. Can revert to green if the roots are disturbed when weeding. Raised by Alan Bloom in 1960s. Height 90cm (3ft) / H4–9

PHLOX X ARENDSII 'LUC'S LILAC'
A tall August-flowering phlox with small heads of luminous lilac flowers. Coen Jansen, probably 1990s. Height 1.2m (4ft) / H3–9

demanding. Pinks tend to be less so. 'Alba Grandiflora' is an airy, tall white phlox that also withstands dry conditions including chalk. White is hard to place under summer sun, however. It can look stark.

Phloxes were widely grown and bred because they make excellent cut flowers. They also follow the roses, so they were often added to rose beds, and their sweetly scented flowers, reminiscent of Madeira cake, are very bee- and butterfly-friendly. They create domed heads of flower which stand out well in the border or vase.

Height 70cm (28in) / H4–9

PLANTING ADVICE
Most border phloxes need good soil. Division is the easiest method. Select the outer pieces and replant in spring, or in September after flowering. You can take root cuttings in winter and they can also be raised from tip cuttings in late spring. Woody clumps should be split. Always deadhead.

'BIG BLUE'

ERYNGIUM X ZABELLI

Spiky electric-blue bracts, etched in white, set around a series of thimbles on a branching framework of stems.

    Deciduous eryngiums, or sea hollies, perform really well in dry summers because they are either tap-rooted or deep-rooted. They all need full sun for most of the day as well. Their spiky flowers persist for many weeks and they keep their colour well in drier summers, but tend to brown following heavy rain.

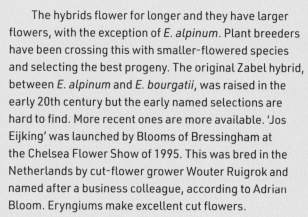

The hybrids flower for longer and they have larger flowers, with the exception of *E. alpinum*. Plant breeders have been crossing this with smaller-flowered species and selecting the best progeny. The original Zabel hybrid, between *E. alpinum* and *E. bourgatii*, was raised in the early 20th century but the early named selections are hard to find. More recent ones are more available. 'Jos Eijking' was launched by Blooms of Bressingham at the Chelsea Flower Show of 1995. This was bred in the Netherlands by cut-flower grower Wouter Ruigrok and named after a business colleague, according to Adrian Bloom. Eryngiums make excellent cut flowers.

Eryngiums are bee magnets and they hybridise and set seeds in the garden setting. 'Blue Waves', a Hillier Chelsea introduction of 2018, was selected from eleven hybrid seedlings found at Dove Cottage Nursery near Sheffield in Yorkshire. It has a branching stem stopped by small blue thimbles surrounded by narrow, jagged bracts. I tend to remove sea holly seedheads, to prevent seedlings. A good clump should produce up to five stems. Height 80cm (32in) / H5-8

FOUR MORE SEA HOLLIES TO TRY

ERYNGIUM GIGANTEUM
The biennial silvery phone-mast version is beloved by bumblebees. Cull most of the stems before the seeds drop, otherwise it will swamp you. Height 1m (39in) / H5-8

ERYNGIUM BOURGATII
A short sea holly with variegated white-veined wispy foliage topped by small blue flowers tinged in green. Self-seeds in paths and crevices. Height 70cm (28in) / H5-8

ERYNGIUM ALPINUM
The largest flowers of all and the only soft, strokable head too, opening green and then colouring up to shades of silver-mauve and cobalt-blue. Good green foliage. Not that easy. Height 75cm (2.5ft) / H5-8

ERYNGIUM PLANUM
Tiny blue flowers, but lots of them, a favourite for cutting. Subtle. Height 75cm (2.5ft) / H5-8

'JACOB CLINE'

MONARDA

This bright-red bergamot, with more-upright florets than usual, is supported by a ring of dark, leafy bracts.

Borders should be about cadence and contrast, to quote the late and great English plantswoman Beth Chatto, and that's what monardas provide. The square stems are topped with a cockade of petals that splay outwards like a shuttlecock. Many come in purples and mauves, but 'Jacob Cline' is a devilish deep-red and it smoulders in the midsummer border just like the heart of a blacksmith's forge. There's an asymmetrical quality to the flowers too, which adds an arrogant air. This stunner was discovered in Georgia, USA, by Don and Jean Cline and named after their garden designer son. It was introduced by Saul Nursery in the USA in the 1980s and it's said to be less susceptible to mildew.

FIVE MORE MONARDAS TO TRY

MONARDA 'SQUAW'
Deep-red flowers and said to be mildew resistant. One of a series bearing Native American names. Height 1m (39in) / H4–9

MONARDA PUNCTATA
Grown as an annual in the UK, but this needs warmth to do well. Lilac whorls of flower followed by a lacy seedhead. Height 60cm (2ft) / H4–9

MONARDA 'BEAUTY OF COBHAM'
Pale-pink flowers held in darker calices. From the 1930s. Height 90cm (3ft) / H4–9

MONARDA 'MARSHALL'S DELIGHT'
Vivid-pink flowers held above dark foliage. Raised in Canada by Henry M. Marshall pre-1980s, for its hardier constitution. Height up to 90cm (3ft) / H4–9

MONARDA 'PRÄRIENACHT'
Rich lilac-purple flowers above dark bracts on a shorter plant. Height up to 80cm (32in) / H4–9

PLANTING ADVICE
Plant in spring in good
soil and try to provide
enough space to allow the
air to flow. This will help
to prevent mildew. Be
prepared to divide every
three years, in spring, to
keep it vigorous.

The standard red is 'Gardenview Scarlet' and this was
selected by Henry Ross of Gardenview Horticultural Park in
Cleveland, Ohio, in the early 1900s. It forms a regular dome and
it's still a great plant, just not quite as sassy as the slightly shorter
'Jacob Cline'. The reds prefer afternoon sun rather than midday
scorch but, after all, they are plants found in damp meadows
and in open woodland. They prefer moister soil than the mauves
and purples, I admit, but few things please the eye as much as
a patch of 'Jacob Cline' in its pomp. You could opt for the softer
mauve, like 'Violet Queen'. It will still flower well, but it's shorter
and it blends into the late-summer border. I prefer explosions!
Unfortunately, I don't get the hummingbird pollinators!
Height up to 1.2m (4ft) / H4-9

'HERRENHAUSEN'

ORIGANUM LAEVIGATUM

The two-tone pink and purple sprays are the best butterfly and bee magnet of all.

Origanum laevigatum is native to Cyprus, Turkey and Syria. However, I've found 'Herrenhausen' completely long-lived and hardy in my cold garden in the Cotswolds. The dark rosette of foliage sends up several wands of flower in the second half of summer and they produce highly concentrated nectar that's appreciated by many August-flying butterflies. Each tiny, cool-pink flower is held inside a purple calyx and, as the flowers fade, they leave a lacy black seedhead. Many aromatic plants classed as herbs also have highly concentrated nectar.

PLANTING ADVICE
Can be planted throughout the year, if pot-raised. Choose a sunny position towards the front of the border and weave several through your planting. Cut back in late autumn and divide in spring – but only if you have to.

The name Herrenhausen comes from a royal garden in Hanover, Germany. The gardens, designed to be a pleasure garden for the royal court, were laid down by the Electress Sophie from 1676 to 1714 and they still cover an area of 135 hectares (334 acres). Another royal garden, Berggarten, literally the 'garden on the hill', was the royal vegetable garden and there is a large-leaved garden sage with this name. Both gardens are still open today.

'Herrenhausen' brightens up the garden in August, a month when many plants look rather tired. It's quite low-level, so it could be used as an edging, or it can grow in the lee of roses preparing for their second flush.
Height 45cm (18in) / H5–9

FIVE MORE AROMATIC PLANTS TO TRY

SALVIA OFFICINALIS 'BERGGARTEN'
This round-leaved sage is larger in every department. The flowers are a soft-blue. Cut back in late spring to keep it compact. Height 60cm (2ft) / H4–9

ORIGANUM VULGARE 'COMPACTUM'
Low-growing mat-forming marjoram with nectar-rich pink flowers. Good in paving or a crevice. Height 20cm (8in) / H5–9

AGASTACHE 'BLUE FORTUNE'
Fuzzy mid-blue bottlebrushes, with a darker core, above nettle-like foliage. A giant hyssop. Height 90cm (3ft) / H4–10

BRONZE FENNEL FOENICULUM VULGARE 'GIANT BRONZE'
Aniseed-scented fine foliage in shades of caramel and toffee form puffs of smoke. Yellow umbels follow. Height 1.5m (5ft) / H4–9

THYMUS SILVER QUEEN'
Lots of clusters of tiny mauve-pink flowers above variegated foliage in cream and green. Pretty and edible. Height 10cm (4in) / H5–9

'ANDENKEN AN FRIEDRICH HAHN'

PENSTEMON

Wine-red slender bells from midsummer onwards.

Penstemons are southern hemisphere plants and they enjoy evenly balanced days and nights. Most of them only get going in the northern hemisphere once the evenings begin to get longer. A hundred years ago wide-flowered penstemons were used as bedding plants in formal Victorian gardenesque schemes. The wallflowers and spring bedding came out and then the penstemons, raised from cuttings, went in. Names like 'White Bedder', a 1912 variety that has masqueraded under many different names including 'Snowstorm', echo their past.

FIVE MORE PENSTEMONS TO TRY

PENSTEMON HETEROPHYLLUS '*HEAVENLY BLUE*'
Earlier to flower, this Californian species has blue flowers tinged with pink-purple. There are many named forms. Height 60cm (2ft) / H7-10

PENSTEMON '*EVELYN*'
A 1934 pale-pink penstemon with small flowers and a sturdy habit. Height 60cm (2ft) / H7-10

PENSTEMON '*STAPLEFORD GEM*'
Exquisite, with two-tone mauve and lilac flowers, heavily veined on the paler inside. Good green foliage. 'Sour Grapes' is similar, with darker flowers. From 1930. Height 60cm (2ft) / H7-10

PENSTEMON '*ALICE HINDLEY*'
Subtle mauve-purple trumpets with neat white throats. Another 1930s stunner. Height 60cm (2ft) / H7-10

PENSTEMON '*BLACKBIRD*'
Ron Sidwell's easiest dark-flowered bird, this is willowy and willing – unlike the miffy 'Raven'. Dates from 1960. Height 60cm (2ft) / H7

PLANTING ADVICE
As with all slightly tender perennials and subshrubs, plant in spring. Deadhead regularly and leave the top growth intact over winter, as frost protection. Cut back to the new basal growth once spring takes hold. Take one or two cuttings from the new shoots as an insurance policy.

Many penstemons are not super hardy and I have lost plenty of them. However, 'Andenken an Friedrich Hahn' is much more resilient. This may be because it has narrower foliage, said to be an indicator of hardiness. Or it may be because it was raised in Switzerland by Herman Wartmann around 1918. He used 'Southgate Gem', a reddish-pink raised in 1910, and selected for hardiness. 'Schoenholzeri' was also raised in Switzerland and both were obtained by Alan Bloom in the 1930s. The Second World War intervened and their German-sounding names were changed to 'Garnet' and 'Firebird', although not by Alan Bloom.

These bee-friendly plants have continued to attract plant breeders and two lived close to Pershore in Worcestershire. Ron Sidwell, a famous plum expert and lover of southern hemisphere genera, raised the Bird series in the 1960s. Edward Wilson raised his Pensham strain from 1985 onwards, naming fifty varieties and selecting for a succession of flowers held on strong stems.
Height 1m (39in) / H7–10

THE PLANTS

87

VAR. DEAMII

RUDBECKIA FULGIDA

The crispest yellow daisy of all, with outward-facing sunshine-yellow petals radiating around a brown-button middle.

It's difficult to know where to lavish praise first. The crisp yellow daisies resist the weather and go on for weeks, bridging the second half of summer and autumn. The stems are stiff, so no staking's required, and the hirsute dark-green foliage is as pristine as the perfectly formed yellow daisies. It's worth remembering that 'daisy' is a corruption of 'day's eye', an indicator that all daisies need a good dose of sunshine every day. This one rewards you with its own version of floral sunshine and it's perfectly poised and elegant, with flowers of just the right

Plant in spring
if possible. No
deadheading is required,
but cut back in autumn
because the heads
absorb water and then
tend to rot. Divide when,
and if, necessary in
spring.

size. Although sometimes called black-eyed Susan, the middles are in fact a rich brown and that suits the late-season colour palette.

I have grown Deam's coneflower in poorer ground quite successfully, even in very hot summers. The natural range extends from New Jersey to Missouri, south to Mississippi and east to Florida, indicating heat and humidity aren't a problem. I consider it drought-tolerant. Hardiness isn't a problem for most either.

There are several similar rudbeckias. 'Goldsturm', which is generally seed-raised, is very similar, but not quite as neat and prim, with slightly floppier petals and less-good foliage. This plant was spotted by Karl Foerster's head gardener, Heinrich Hagemann, in Graz Botanic Garden, having been raised from American collected seeds. He passed the plant to Foerster in 1938 and Foerster named it 'Goldsturm', meaning 'gold storm', when he saw it in flower. Both plants take low-maintenance to a new level.
Height 90cm (3ft) / H3–10

FOUR MORE RUDBECKIAS TO TRY

RUDBECKIA FULGIDA VAR. FULGIDA

This flowers until late autumn, into November, extending the yellow-daisy season and it's a more open plant. Height up to 90cm (3ft) / H3–10

RUDBECKIA SUBTOMENTOSA 'HENRY EILERS'

Tall and columnar with spoon-billed yellow petals surrounding a brown middle. Found beside a railway in Montgomery County, Illinois, in the USA, by a retired nurseryman of the same name. Height 1.2m (4ft) / H3–10

RUDBECKIA HIRTA 'INDIAN SUMMER'

Although considered annuals or biennials, I have found some come back for longer. Lots to choose from, but this has enormous yellow flowers! Height 90cm (3ft) / H3–9

RUDBECKIA FULGIDA 'LITTLE GOLDSTAR'

A small version for those who like short! A Jelitto introduction from 2011. Height up to 40cm (16in) H3–10

'MATRONA'

HYLOTELEPHIUM

Fleshy, large leaves, the colour of a pigeon's wing, and upright stems topped by dusky pink domes.

Many of these larger sedums were raised in Germany and the forerunner was Georg Arends' 'Herbstfreude', re-named in post-war Britain as 'Autumn Joy'. This, along with 'Ruby Glow', was introduced into Britain by Alan Bloom, who admired Georg Arends. Many more have followed including the Dutch-bred 'Xenox' and Ernst Pagels's 'Karfunkelstein'. The latter two are very similar, but the younger foliage of 'Karfunkelstein' is darker than that of 'Xenox'.

'Matrona' is thought to be a hybrid between 'Spectabile', a glaucous-leaved pink-flowered parent of 'Herbstfreude', and an upright handsome maroon-red sedum named *Hylotelephium telephium* (Atropurpureum Group). It was selected from a batch

FOUR MORE SEDUMS TO TRY

HYLOTELEPHIUM TELEPHIUM (ATROPURPUREUM GROUP)

This airy plant reaches for the sky with outstretched arms, quite unlike any other sedum. The stiff stems divide into slender, upward-facing branches, topped by small, rounded clusters of maroon flowers. Height 50cm (20in) / H3–9

HYLOTELEPHIUM 'RED CAULI'

Found at Marchants Plants in 2000, the intense red flowers have a hummocked and bumped shape – just like cauliflower. Height 50cm (20in) / H3–9

HYLOTELEPHIUM 'VERA JAMESON'

A front-of-border sprawler, for a sunny edge, with purple-black foliage and fuzzy pink flowers. Joe Elliott, 1970. Height 15cm (6in) / H4–9

HYLOTELEPHIUM 'CARL'

Introduced by Joe Sharman of Monksilver Nursery in Cambridgeshire, this blue-green leaved sedum bears lots of rose-pink smaller heads. Height up to 60cm (2ft) H3–9

PLANTING ADVICE
These are plants in need
of a sunny position. They
benefit from regular
division, because they can
get woody in the middle.

of seedlings in 1991 by Ewald Hugin, a plantsman from Freiburg,
Germany, according to Beth Chatto's website. 'Matrona' has
inherited the glaucous foliage of 'Spectabile', with a purplish flush
from Atropurpureum. The pink flowers are domed and the whole
thing is upright, so this stately plant provides the perfect contrast
to all those mound-forming perennials.

'Matrona' translates as 'dignified married woman', which
hints at a rounded matronly figure. The fleshy drought-tolerant
foliage endures and wide heads of dusky flowers fade to shades
of chocolate. As a result, 'Matrona' give weeks of interest without
any tedious preening. Most hybrid sedums usually occur naturally
after pollination. 'Purple Emperor', 'Red Cauli' and 'Marchants
Best Red' were all seedlings named by Graham Gough of
Marchants Plants in the UK. Terra Nova's introductions, 'Carl',
'Cloud Walker' and 'Mr Goodbud', were given the RHS's Award of
Garden Merit in 2006.
Height 60cm (2ft) / H3-9

'ALBUM'

VERBASCUM CHAIXII

Dainty spires of small white flowers, each one with a damson middle punctuated by orange stamens.

Most verbascums are short-lived and difficult to keep from one winter to the next. However, this clump-forming European verbascum endures from year to year, putting up slender tapers of white flower on dark stems once summer arrives. The grey-green leaves are a bonus too, but it's the combination of white petals blotched in damson, plus the orange stamens, that make this so easy to use in a border. It will go with so many plants, whether it's picking up dark roses or bright-pink cosmos. And it will echo anything that's peach or orange, due to the anthers. The flowers never look stark, despite being white, so they sit happily in the border without jumping out, as a clear-white would.

This verbascum is easily raised from seed, although care must be taken to get seeds marked 'Alba', for the ordinary species has small yellow flowers. It's worth running this through a border,

FOUR MORE SPIRES TO TRY

TEUCRIUM HIRCANICUM

Sage-like foliage and deep-purple tapers of flower in late summer. Good winter seedheads. Height 60cm (2ft) / H5–8

ANTIRRHINUM MAJUS 'NIGHT AND DAY'

Dark foliage and deep-maroon and white flowers on this heritage snapdragon. Height 45cm (18in) / H4–10

ALCEA ROSEA - HOLLYHOCK

Go tall, when you opt for a hollyhock, because the squat ones are awful. Ignore the darkest and go for something luminescent like 'Halo Apricot'. Height on average 1.5m (5ft) / H3–9

MALVA SYLVESTRIS 'ZEBRINA'

A mallow with pale-pink flowers zebra-striped in deep-purple. Height 90cm (3ft) or more / H7–11

to give a vertical presence. You could equally well sow seeds of *V. phoeniceum*, an early summer-flowering verbascum with purple, pink or white flowers. This has a winter rosette, so it has a head start, making an early impression in a summer border. It will self-seed too.

Verbascum blattaria f. *albiflorum* can also be grown from seed and this is another biennial, making a rosette in the first year and then flowering in the second. The white flowers, well-spaced along the stem, have a pink colourwash and the middle of the flower is a deep-purple tangle of stamens and style. Let it self-seed. Lay the flowering spikes down if you have to. I find verbascums fascinating at every stage. The square buds look like damask embroidery, before the bee-friendly flowers open. Height 90cm (3ft) / H5–9

PLANTING ADVICE
These all like well-drained soil and a good measure of sun. *V. chaixii* 'Album' is long-lived, but the ones that produce copious amounts of seeds are either biennial or very short-lived. Save your own seeds.

'ORION'

GERANIUM

Perpetually flowering single cobalt-blue saucers, veined in dark-red with a hint of green and white in the middle.

Hardy geraniums, or cranesbills, are members of the plant chorus line. They provide the backing track for showier divas like fully double peonies, roses and oriental poppies. There are over 260 hardy geranium species in cultivation and when two closely allied species hybridise, following pollination, the progeny are sometimes sterile. They are unable to produce seeds, so they flower on and on. 'Orion' is one of these sterile, spontaneous

'Orion' does not like cold, waterlogged soil, but it will grow in partial shade or full sun. Well-grown specimens can be planted during the growing season. Divide every four to five years in April.

hybrids and it came to prominence during the RHS hardy geranium trials held at RHS Wisley between 2002 and 2007.

'Orion' was not known to the specialist judging panel at that time, but it soon outshone the benchmark 'Johnson's Blue' because it produced non-stop flowers from May until late. They were larger than any other single blue hardy geranium. It arrived on the trial almost by accident. It was discovered in the late 1990s by a Dutch nurseryman named Brian Kabbes. Another Dutch nurseryman, Coen Jansen, volunteered to submit 'Orion' to the RHS hardy geranium trial. It's thought to be a natural hybrid between 'Brookside' and *G. himalayense* 'Plenum'. It produces a fantastic flush between May and late June and then the flowers dwindle. That's the moment to cut it right back to zero. Within a week new foliage appears and then more flowers come and continue into autumn. You won't get unwanted seedlings from 'Orion' and that's a blessing.
Height 60cm (2ft) / H3–9

FOUR MORE LONG-FLOWERING HARDY GERANIUMS TO TRY

GERANIUM 'BALLERINA'

Alan Bloom's front-of-a-sunny-border hardy geranium, from 1962, has pale-lilac flowers heavily veined in purple with a red-purple middle. Height 15cm (6in) / H5–9

GERANIUM HIMALAYENSE 'DERRICK COOK'

Large grey-white flowers finely veined in purple, above green foliage. Collected by Derrick (of Red Gables garden in Worcestershire) in 1984. Height 30cm (12in) / H5–9

GERANIUM 'IVAN'

From Ivan Louette of Belgium, this *G. psilostemon* hybrid has very large flowers on a large plant. There are no gaps between the petals on Ivan's rounded flowers. Height 60cm (2ft) / H5–6

GERANIUM 'JOY'

Evergreen marbled foliage and pale-pink silky flowers, with dark-reddish veins. From Alan Bremner. Height 40cm (16in) / H5–6

'SEAL'

LAVANDULA X INTERMEDIA

Tapering spires, in clear blue-purple, in the second half of summer.
Lavenders vary in hardiness, flowering times and in the shape
of the flowering spike. They are all sun-loving aromatic subshrubs.
'Seal' is a 1930s lavandin and these are all hybrids between English
lavender *Lavandula angustifolia* and spike lavender or *L. latifolia*.
Lavandins flower in July and August and their flowers are slender,
tapering to a point at the tip. They are not blocky, because their
willowy, branching stems make up two-thirds of their height. Their
leaves make up the lower third, so they're very graceful.

Lavandins float like clouds, swaying and moving in the slightest
breeze. They can reach a metre in height, so they are best given a
prominent position in the garden, at the end of a sunny border for
instance. Lavandins need a careful pruning regime, because they

FOUR MORE LAVENDERS TO TRY

L. ANGUSTIFOLIA 'MELISSA LILAC'
A soft-lilac, June-flowering,
English hardy lavender with
larger florets and greyer
foliage. Makes a good
specimen. Height 45cm
(18in) / H5-10

LAVANDULA STOECHAS SUBSP. *PEDUNCULATA* SPANISH LAVENDER
A flamboyant lavender with
long, sterile bracts like ears,
these wave about. Needs a hot
spot and tends to be short-
lived. Height up to 90cm
(3ft) / H7-9

LAVANDULA STOECHAS SUBSP. *STOECHAS* FRENCH LAVENDER
Shorter tufts on top of the
flowers and a squat, sturdy
habit. Height 80cm (32in) /
H7-9

LAVANDULA 'WILLOW VALE'
There has been a lot of
breeding in New Zealand
and Australia. 'Willow Vale'
raised in Australia in 1992, has
crinkly-eared purple flowers
with paler purple ears. Height
60cm (2ft) / H7-9

are hardy to -15°C (5°F). The foliage is trimmed and shaped back to the top of the main leafy growth in early September, just as the flowers begin to fade. They often make good winter silhouettes, because they keep a billowing, rounded shape. Hard pruning will result in frost or rain damage. Given the right conditions they can last for roughly ten years.

Lavandins are highly aromatic and they are grown in the fields of Provence for their oil, which is used in detergents, air fresheners and polishes. The classic oil variety is 'Grosso'. The neat, purple flowers fan out from a mass of bright-green foliage. The Dutch Group have a more upright stance, with pale-purple stems rising vertically above greyer foliage.
Height up to 90cm (3ft) / H5–8

PLANTING ADVICE
Try to plant in the spring and be prepared to water in dry weather during the first growing season, because the deep root system they rely on is not yet formed. English lavender, *L. angustifolia*, is short and is perfect for hedging. The flowers consist of short blunt-tipped spikes in June. It can be pruned back hard in August. Lavenders with tufted flowers are the least hardy of all, but they do produce a succession of flowers between May and September.

'HONORINE JOBERT' JAPANESE ANEMONE

ANEMONE X HYBRIDA

Perfectly round single white flowers with a boss of yellow stamens set around a green pincushion middle.

Japanese anemones ramble and that puts some gardeners off. Personally, I love their ability to pop up and fill the gaps among roses and peonies. They extend the planting in summer-themed herbaceous borders, because these long-stemmed plants perform after the main summer flush but before the late-late set get going. It was their ability to ramble and pop up that drew the attention of Scottish plant hunter Robert Fortune (1812–1880). While in Shanghai, he spotted Japanese anemones rambling through a graveyard.

PLANTING ADVICE
Like many plants with a wandering nature, Japanese anemones can be difficult to propagate and get going. Start with a well-grown nursery plant in spring. If you do pot up runners, put them into good compost. They do not like waterlogged soil, but many will tolerate drier shady areas.

Anemone hupehensis was introduced into Europe in 1844, but it became known as the Japanese anemone, because it has been grown in Japanese gardens for centuries. It's native to Hupeh province in China.

Most Japanese anemones come in shades of pink and their habits vary from 'slowly wandering' to 'I'm going to take over the world'. The arrival of this well-behaved pure-white is hard to explain, because Japanese anemones rarely set seed. Perhaps it was a sport, a shoot from a pink-flowered plant. We'll never know, but this superb Japanese anemone has been around since 1858, some 160 odd years. It was found in Honorine Jobert's garden in 1858 and the French nurseryman Victor Lemoine (1823–1911) recognised its elegance and launched it. I have often called it the Audrey Hepburn of the plant world.

It loves a little shade and, being a pure-white, looks better for it. The foliage is low and green. There are other inferior whites, including the ragged 'Whirlwind', but none capture the simplicity and elegance of 'Honorine Jobert'. She doesn't ramble far. Height up to 1.2m (4ft) / H5–8

THREE MORE JAPANESE ANEMONES TO TRY

ANEMONE HUPEHENSIS 'HADSPEN ABUNDANCE'

Single pink flowers with petals of uneven size. The two opposite petals are often a brighter pink, so this appears to dazzle in the border. 'Bowles's Pink' is a lookalike and, dare I say, a better plant. Height 80cm (32in) / H5–8

ANEMONE HUPEHENSIS VAR. JAPONICA 'PRINZ HEINRICH'

Another oldie from 1902, with slightly less neat semi-double pink flowers containing slightly quilled petals.

It flowers for longer than all the rest. When trialled at the Chicago Botanic Garden, it flowered for 64 days. Height 90cm (3ft) / H5–8

ANEMONE X HYBRIDA 'KÖNIGIN CHARLOTTE'

The frillier than usual lower leaves overlap each other, like piled up scallop shells, and the flowers have neatly arranged pale-pink petals with darker pink backs. Raised by Wilhelm Pfitzer in 1898. Vigorous. Height 1.2m (4ft) / H5–8

'FERNER OSTEN'

MISCANTHUS SINENSIS

Early plumes of maroon-red grassy heads fading to mink-brown.
These late-season grasses make the best winter silhouettes
of all and there's plenty of variety. Your geographical location
affects flowering times and this, in turn, dictates whether nuisance
seedlings appear in your garden. In the heart of England, where my
garden is, the earliest miscanthus is always 'Silberfeder' and that
only produces flowering plumes in early September. Go 100km
(60 miles) west or south, and it could flower six weeks earlier.
'Ferner Osten', meaning 'Far East', raised by German nurseryman
Ernst Pagels (1913–2007), bears plummy-red plumes with a
coppery sheen on a compact plant up to 1.5m (5ft) in height. The
plumes should bear a white tip.

Pagels endured colder winters than I do, so he had to place
'Gracillimus' in a greenhouse in order to persuade it to set seeds.

FOUR MORE LATE-SEASON GRASSES TO TRY

MISCANTHUS SINENSIS 'SILBERFEDER'
Dismissed by many
nurserymen because it
flowers first, but this majestic
miscanthus performs reliably
in August even in cold districts.
Substantial silver plumes.
Height 1.8m (6ft) / H4–9

MISCANTHUS SINENSIS 'MALEPARTUS'
Upright purple-red flowers
and broad leaves that turn
pale-brown in autumn – many
people's favourite. Height 2m
(6.5ft) / H4–9

MISCANTHUS SINENSIS VAR. CONDENSATUS 'COSMOPOLITAN'
A spectacular foliage variety,
with green leaves middled in
cream. It wanders, rather than
making a tight clump. Height
up to 2m (6.5ft) / H4–9

CORTADERIA SELLOANA 'PUMILA'
A compact Pampas grass. The
plumes appear in autumn and
last through winter, forming
exclamation marks in the
border. Height 1.5m (5ft) /
H8–10

The first batch of seedlings showed promise and variation and this prompted Pagels to breed and name at least twenty-five from the 1950s onwards. When RHS Wisley held their miscanthus trial from 1998 to 2003, AGMs were given to 'Kleine Silberspinne', 'Flamingo', 'Gewitterwolke', 'Ghana', 'Undine', 'Kaskade', 'Grosse Fontäne', 'Septemberrot' and 'Kleine Fontäne' as well as 'Ferner Osten'. Their grassy heads, which range from silver through to plum, endure through winter to late spring.

Miscanthus was grown for its foliage by two eminent gardening pioneers. Gertrude Jekyll grew 'Zebrinus' at Munstead Wood. She used the horizontally banded green and gold foliage to create light and shade. William Robinson (1838–1935), of Gravetye Manor, also grew miscanthus under the name of *Eulalia japonica*, or the Chinese silver grass.
Height 1.5m (5ft) / H4–9

PLANTING ADVICE
Use a specialist grass nursery, if possible. Newly bought miscanthus take at least three years to clump up to a decent size and many specialist grass nurseries recommend planting in threes and fives or by the square metre. Miscanthus do best in good soil and they like summer moisture. Planting throughout the growing season is fine. Cut back in early spring, because they reshoot early.

'ARENDSII'

ACONITUM CARMICHAELII

A September spire of hooded cobalt-blue flowers held on self-supporting downy stems.

Blue is a rare colour in the garden and this stately monkshood is a stunner in the clear light of September, a month of mound-forming golden yellow and purple daisies. Nurserymen have tended to avoid them, most probably due to their toxicity. I repeat, care must be taken and gloves and sleeves are the order of the day when handling this. However, these later aconitums make great garden plants. They avoid the ragged foliage of the earlier flowering ones. They are handsome and I can clearly recall the effect of seeing 'Arendsii' planted behind miscanthus in Alan Bloom's Dell Garden, long before tall grasses became fashionable. Georg Arends (1863–1952) hybridised perennials including astilbes, sedums, hardy chrysanthemums and hostas. The nursery was almost destroyed during the Second World War, although it was rebuilt. After his death, most of the site went for housing.

THREE MORE RICH BLUES TO TRY

AGAPANTHUS 'NORTHERN STAR'

Dick Fulcher's Devon-bred, true-blue deciduous agapanthus. Lots of flowers and dark shading on the foliage base. Feed and water well. Height up to 80cm (32in) / H7–9

SALVIA 'AMISTAD'

Most tall, deep-blue tender salvias flower late in the year: this produces flower spikes throughout summer and autumn. Found by Argentinian Rolando Uria in a garden centre. Amistad means 'friendship'. Height 1.2m (4ft) / H up to 8

PULMONARIA 'BLUE ENSIGN'

Rough green leaves, with no spotting, topped by deep-blue flowers in spring. Needs regular division. Spotted in 1990 in Bowles Corner at RHS Wisley. Height 25cm (10in) / H3–9

Good soil is appreciated, but
their woodland provenance
allows them to grow close
to shrubs and at the back of
the border. 'Arendsii' must
be deadheaded.

His aconitum, named 'Arendsii', is the very best. It has glossy
foliage, thick downy stems and a short spire containing fifty or
so flowers. Green seedheads follow on and it's important to
deadhead this at all costs, for the seedlings will infiltrate the true
'Arendsii'. Inferior plants, raised from seed, are known as the
Arendsii Group. Seek out the true 'Arendsii', from a good nursery,
because it's one of September's highlights. It offers a unique
profile, a glowing colour, and it will tolerate some shade too.
Height 1.5m (5ft) / H3–9

'CANGSHAN CRANBERRY'

SANGUISORBA

Deep-red bobbles, held on tall stems, appear in autumn before fading into winter like a swarm of bees.

Sanguisorbas, or salad burnets, are overlooked in gardens, but there's a lot of variety because their geographical range stretches from North America, eastwards into China. The Asian plants flower later in the year and they tend to be tall and quite thirsty. Flower shape varies, according to the pollinator. *S. hakusanensis*, the Korean mountain burnet, has grey foliage and catkin-like bright pink flowers that are especially favoured by butterflies. 'Lilac Squirrel'

PLANTING ADVICE
Damp meadows are the natural home of most, but they thrive in garden soil and are able to push up through other plants. Deadheading is advisable, particularly with *Agalinis tenuifolia*. Deep roots.

is a named form with long catkin-like pink flowers. It's a standout flower, but difficult to place.

The later ones usually have bobble-shaped red flowers that smell unpleasant. Their musky smell is often described as proteinaceous, a polite way of describing the smell of male emissions. This scent attracts small flies and most sanguisorbas produce seedlings. This has led to a rash of similar ones in recent years. Most are airy, with stems that rise well above the foliage and they could be mistaken for a grass from a distance. They provide flowing movement and have an elegant willowy presence, so they mix well with grasses in prairie-style planting. Piet Oudolf's 'Arnhem' was the forerunner, but it flowered in summer before the grasses and taller herbaceous got going.

The most useful sanguisorbas give late, bobble-like flowers in maroon-red. 'Cangshan Cranberry' is less see-through than most, because the foliage is substantial. The bobbles are long ovals in shades of cranberry-pink. It was collected in a wet meadow in Yunnan by Marina Christopher of Phoenix Perennial Plants. It peaks in September but carries on until November.
Height 2m (6.5ft) / H4–8

THREE MORE SANGUISORBAS TO TRY

SANGUISORBA OFFICINALIS 'MARTIN'S MULBERRY'

This arrived as a seedling from Martin Lowne of West Acre Gardens. It forms a large clump with wine-purple thimbles on airy stems in August – for me. Possibly earlier for you. Height 1.8m (6ft) / H4–8

SANGUISORBA 'TANNA'

A short sanguisorba with claret-red, slender heads on dark reddish stems. These glow against small, deep-green serrated leaves. Flowers from June until August, picking up the purple veins and dark stamens of hardy blue geraniums such as 'Orion'. Height 40cm (16in) / H4–8

SANGUISORBA OFFICINALIS 'RED THUNDER'

A late-summer performer with deep-red flowers held on stiff green stems. This Korean burnet is a Piet Oudolf favourite and he may have named it. Height 1.2m (4ft) / H4–8

'LITTLE CARLOW'

SYMPHYOTRICHUM

A cloud of tiny, intense blue flowers, encased in small diamond-patterned buds in early September.

If you were able to walk along a border of asters, I can guarantee you'd stop and stare at this one because it stands out from all the others. It's easy to grow and the colour, deepened by autumnal light, seems bluer than any other aster. The small leaves can be almost hidden under the panicles of lavender-blue flowers and, if this aster's happy, it will produce a substantial clump. This aster was one of several raised by Mrs Thornley in Devizes in Wiltshire between 1930 and 1940. 'Little Carlow' inherits its flowering habit from *Symphyotrichum cordifolium*, the Blue Wood aster. The deeper lavender-blue flowers are inherited from an unknown New York aster, *S. novi-belgii*.

FOUR MORE AUTUMN-FLOWERING ASTERS TO TRY

SYMPHYOTRICHUM NOVAE-ANGLIAE 'HELEN PICTON'

A cross between two New England asters, the bright-pink 'Andenken an Alma Pötschke' and 'Purple Dome', this is one of the few deep-purple easily grown asters. Bred by Paul Picton in the UK. Height up to 1.3m (4.2ft) / H4–8

SYMPHYOTRICHUM NOVAE-ANGLIAE 'JAMES RITCHIE'

Cerise-pink flowers with long-lashed petals set around a golden middle. Height 1.2m (4ft) / H4–8

SYMPHYOTRICHUM ERICOIDES 'PINK CLOUD'

Masses of small purple-pink flowers in October, on a bushy plant. Good disease resistance. Height up to 90cm (3ft) / H4–8

EURYBIA DIVARICATA

This easily grown aster has small white flowers in August, but I love the black wavering stems and diamond-shaped green leaves. Good around box. Height 90cm (3ft) / H4–8

PLANTING ADVICE
All autumn-flowering
plants should be divided
in the spring, just as they
break into growth, but only
if they need it. They like
sun and reasonable soil
and they cut well. Remove
the spent flower stems
in autumn, to prevent
unwanted seedlings. Some
asters run.

'Little Carlow' has endured for almost a hundred years,
because it's easy to grow. This cannot be said for all of them. New
York asters, for instance, are high-maintenance plants because
they're found naturally in wet meadows in eastern Canada and
the eastern United States. Water stress leads to mildew in the
garden setting and they also need regular division. They were
highly bred, because their flowers came in desirable shades of
blue. Ernest Ballard, of Old Court Nurseries in Worcestershire
in the UK, raised hundreds of them in the early years of the 20th
century. They included the double blue 'Marie Ballard'.

New England asters, *Symphyotrichum novae-angliae*,
are tolerant of drier conditions and rarely need division. Their
only failing is shabby foliage on their lower stems. They need to
be camouflaged behind another perennial or aster, such as
'Little Carlow'.
Height 1.2m (4ft) / H4-8

'HERBSTSONNE' (AUTUMN SUN)

RUDBECKIA LACINIATA

Tall yellow daisies with swooning petals set around a cool-green cone.

Autumn is a season of tall perennials in warm colours and this is one of the best tall plants, because the daisies are held aloft from the foliage on strong stems. It's clump-forming and a lot of tall yellow daisies aren't. They have foot-shaped roots and they – to paraphrase that Nancy Sinatra song – will walk all over you. This one forms a tight clump and maintains a columnar shape which allows you to weave it through a border or create a backbone.

FIVE MORE DAISIES TO TRY

SYMPHYOTRICHUM LAEVE 'CALLIOPE'

Tall lavender-flowered aster with dark foliage and shiny dark stems. Never needs staking. Airy and reliable. Height 1.2m (4ft) / H4–7

HELIANTHUS GIGANTEUS 'SHEILA'S SUNSHINE'

This roams and sends up the occasional extra tall stem bearing several lemon-yellow flowers, each one with a dark middle. Height 2m (6.5ft) / H4–9

LEUCANTHEMELLA SEROTINA

Another rambler, bearing strong, tall stems topped with white single daisies late into autumn. They face you full on, so it's good against a sultry background, such as *Cotinus coggygria* 'Royal Purple'. It spreads, though! Height 2m (6.5ft) / H4–9

BIDENS HETEROPHYLLA

This short-lived perennial will run in poor soil so it is hit and miss. If it works, it provides delicate yellow daisies on wiry stems. Height 50cm (20in) or taller / H9–10

CHRYSANTHEMUM 'E.H. WILSON'

Tiny, but highly scented cream to yellow pompoms late into the year. Height 75cm (2.5ft) / H3–9

'Herbstsonne' is always listed under *R. laciniata*, but expert opinion believes it to be a hybrid between *R. laciniata* and *R. nitida*. It doesn't self-seed, so it probably is a sterile hybrid, and the flowers last for many weeks. The cut-leaved foliage stays low, so these daisies appear to pop up from nowhere. But it's the green cone that makes them so irresistible to look at. Being a man-high plant, almost on stilts, means that it must have equally tall companions to avoid doing a 'little and large' act. It will flower in some shade. This upward-facing daisy doesn't chase the sun, so you get a good view. Think spinning plates!
Height 1.5m (5ft) / H3–10

PLANTING ADVICE
Divide in spring, a golden rule for all late-flowering plants. The young foliage can be prone to damage from slugs and snails in wet summers. Deadhead in late autumn.

ATROPURPUREUM GROUP

EUTROCHIUM MACULATUM

Whorls of green leaves on almost black stems sit below late-summer storm clouds.

    The recent name change is confusing and most gardeners will know it as *eupatorium* or spotted Joe Pye weed. Whatever the name, this butterfly magnet adds drama to the late-summer border. It needs moisture early on to get going, so I hesitated slightly before adding it to my tough plants list. Should it miss a year, as it sometimes will following a dry spring, it comes back

Plant more than one, in spring if possible, and water well in the first growing season. Cut back in autumn and, if you need to divide, do so in spring.

fighting in the next, so in it goes. It has such a presence, due to those stiff, dark stems, but it's the cloud-effect from the many rosy-purple to maroon globular heads that make this so different.

It's traditionally grown at the back of the border, but I'd like everyone to shuffle it forward, because it shines for three months running. I don't get seedlings, but you may. Nothing else provides that moody infusion of late colour, hovering between red and purple. The pointed buds open gradually to form spidery collections of flowers. It's the fuzziness I love and it chimes with late asters, yellow daisies and beige-tinted tall grasses. Height up to 1.5m (5ft) – perhaps more on good soil / H3–8

FIVE MORE 'DOMES' TO TRY

HYDRANGEA ARBORESCENS 'ANNABELLE'

An undemanding hydrangea that can cope with cold winters and hot summers. Discovered growing wild near Anna in Ohio, hence the name, and launched in the mid-1970s by the Gulf Stream Nursery. The white flowers have a green immature stage and they fade to pink. Height 1.2m (4ft) / H3–8

SELINUM WALLICHIANUM

The 'queen of the umbellifers', according to E.A. Bowles, with lacy green leaves, pinkish-purple sheaths and stems, before cream-white flowers open. Height 1.2m (4ft) – more on good soil / H4–7

BUPLEURUM FRUTICOSUM

One of the few shrubby bupleurums, with evergreen foliage topped by lime-green umbels from midsummer. Give it space to shine. Height 2m (6.5ft) / H6–10

HYLOTELEPHIUM TELEPHIUM 'MATRONA'

The dusky pink flowers pick up the darker bupleurum heads. Height 90cm (3ft) / H4–7

ACHILLEA FILIPENDULINA 'GOLD PLATE'

Gold domes on this tall, stiff-stemmed achillea. These dry well too. Needs good drainage. Height up to 1.8m (6ft) / H3–8

PART 3
PLANTING PARTNERS

Star plants need good partners, ones that enjoy the same conditions and perform in the same season. Here are four for each of the forty tough plants in Part 2. They have been chosen for their ability to perform well for a wide range of gardeners.

This section is designed to help gardeners and garden designers who want to select and combine excellent plants. Use them in threes, fives, sevens or nines and avoid making blobby shapes. Weave ribbon-shaped drifts together and then they'll make more of an impact on the eye. Happy gardening!

BELOW
Prunus 'Kursar'.

'ROBINSONIANA' ANEMONE NEMOROSA

ERYTHRONIUM CALIFORNICUM 'WHITE BEAUTY'

White Tiffany lamps unfurl above beautifully mottled leaves. Cheap and easy! Height 20cm (8in) / H3–9

HYBRID HELLEBORES (HELLEBORUS X HYBRIDUS)

Long-lived perennials come in a range of colours and forms, but the paler singles make the greatest impact visually. Height 30–35cm (12–14in) / H6–8

BRITISH BLUEBELLS (HYACINTHOIDES NON-SCRIPTA)

Cobalt-blue flowers hang from one side of the stem and add a graceful presence to the woodland border. Height 30cm (12in) / H4–9

SIBERIAN SQUILL (SCILLA SIBERICA)

The best self-seeding blue bulb for shadier spots, so no deadheading. It will mingle with all sorts of woodlanders. Cheap and easy! Height 15cm (6in) / H4–9

'JACK FROST' TERRA NOVA BRUNNERA MACROPHYLLA

PRIMULA 'SUNSHINE SUSIE'

A Barnhaven double that lasts well in gardens. The apricot-yellow flowers catch the mood of spring. Height 15cm (6in) / H4–9

SILENE FIMBRIATA

A white campion with airy wands of white-fringed flowers. Tuck it up in shade, even dry shade. Height 90cm (3ft) / H5–9

X HEUCHERELLA 'TAPESTRY'

The three-lobed, red-veined leaves are colourwashed in green and silver and deep-pink flowers follow on. Height up to 30cm (12in) / H4–9

POLYGONATUM X HYBRIDUM

Solomon's seal is architecturally pleasing, although the stems emerge late. Mark them well, or you may trample them before they come up. Height 120cm (4ft) / H5–9

'HYBRID HELLEBORE' HELLEBORUS X HYBRIDUS

PRUNUS 'KURSAR'

The perfect pink cherry for a smaller garden: it provides light shade and all-important drainage. Height 3m (10ft) in 20 years / H5–7

PRUNUS INCISA 'KOJO-NO-MAI'

A spring snow flurry of delicate blush-white flowers held on divaricate, twisted branches. More shrub than tree and slow-growing. Height 2.5m (8ft) / H5–7

MILIUM EFFUSUM 'AUREUM'

Use this golden grass with dark hellebores, just pluck out any unwanted seedlings. Height 40cm (16in) / H5–8

MUSCARI AZUREUM

A non-invasive grape hyacinth with tightly packed pale-blue flowers that open from the bottom of the flowering spike upwards. Height 15cm (6in) / H5–8

EUPHORBIA EPITHYMOIDES

GEUM 'BELL BANK'
Nodding, woodland geum found in the
late Geoffrey Smith's Yorkshire garden,
with larger frilly-edged, red-pink flowers.
Height 40cm (16in) / H5–9

ALLIUM 'PURPLE RAIN'
Later and shorter than 'Purple Sensation',
with softer-coloured mauve flowers. Height
85cm (34in) / H3–8

SCILLA MISCHTSCHENKOANA
Grey-blue ground-hugging hardy scilla that
appears with the snowdrops and cyclamen.
Grow it, even if you can't pronounce it!
Height 10cm (4in) / H4–8

SCILLA FORBESII 'BLUE GIANT' (PREVIOUSLY CHIONODOXA)
Undergoing a botanical shuffle, but 'Blue
Giant' has larger blue flowers than the
species with a more noticeable central
white throat. Self-seeds in brighter
positions. Height 15cm (6in) / H4–8

'TREVI FOUNTAIN' PULMONARIA

HELLEBORUS X HYBRIDUS – YELLOWS AND GREENS
Blue pulmonarias shine close to yellow and
lime-green hellebores. Height 35cm
(14in) / H6–8

PRIMULA 'FRANCISCA'
This Canadian primrose, discovered by
Francisca Darts on a traffic island, has lime-
green ruffled-edged petals. Height 15cm
(6in) / H3–8

GERANIUM PHAEUM 'JOSEPH GREEN'
Wands of double-purple flowers, like
small pompoms, tipped with lime-green
stamens. Found by Lynne Edwards in
2013 and named after her father, a keen
horticulturist. It goes on and on, because
it's sterile. Height 60cm (2ft) / H3–9

GERANIUM MACRORRHIZUM 'WHITE NESS'
A smaller, more compact hardy geranium
with white flowers held above grass-green
foliage. Height 30cm (12in) / H4–9

'RICHARD KAYSE' POLYPODIUM CAMBRICUM

HAMAMELIS X INTERMEDIA 'AURORA'
This butterscotch-coloured witch hazel
has a freesia scent and it's possible to
plant right up to the trunk, as with all witch
hazels. Height 3.5m (11.5ft) / H3–8

DAPHNE LAUREOLA
A British native evergreen with high-gloss
green leaves topped with lime-green,
slightly fragrant flowers. Tolerates deep
shade, but self-seeds. Height 1.2m (4ft) / H6

EPIMEDIUM 'SPINE TINGLER'

Grown for its arrow-shaped green leaves. They have crisply serrated edges and they're held on slender stems so the leaves dance. Height 35cm (14in) / H5-9

SARCOCOCCA CONFUSA

A sweetly fragrant evergreen with highly scented ivory-white flowers set against shiny rich-green foliage. Height up to 1.5m (5ft) / H6-9

'GREEN SPICE' HEUCHERA

X HEUCHERELLA 'KIMONO'

This enduring 1999 hybrid between *Heuchera* 'Green Spice' and *Tiarella* 'Pink Pendant' provides a metallic mix of purple, green and silver foliage. Height 25cm (10in) / H4-9

TIARELLA WHERRYI 'WHERRY'S FOAM FLOWER'

Delicate in looks, but tough in reality, with green lobed leaves and black, wiry stems topped with foamy blush-white flowers in spring. Height 25cm (10in) / H3-9

COTINUS COGGYGRIA 'ROYAL PURPLE'

Woodland plants need shade providers and this wine-red smoke bush creates a contrast. It comes into leaf in the second half of April, but the leaves linger into the second half of November. Height 4-5m (13-16.5ft) / H5-8

LAMIUM MACULATUM 'BEACON SILVER'

A ground cover plant with silver foliage, edged in green and very early bee-pleasing, purple-pink flowers. Height 25cm (10in) / H3-9

'MOONSHINE' ACHILLEA

POLEMONIUM 'LAMBROOK MAUVE'

This soft-mauve Jacob's ladder, which came from Margery Fish's Somerset home, doesn't produce seedlings. The soft-orange centre of each flower adds extra luminosity. Height 45cm (18in) / H4-8

SALVIA 'BLUE SPIRE' - RUSSIAN SAGE - PREVIOUSLY PEROVSKIA ATRIPLICIFOLIA

A white-stemmed, upturned besom brush clothed in small lavender-blue flowers. Hardy and able to cope with moist soil. Height 1.2m (4ft) / H4-9

ALLIUM 'GLADIATOR'

A solid sphere of lilac-purple, star-shaped florets on this long-flowering allium. Weave through to create ribbon-like strips. Height up to 1m (39in) / H4-9

NEPETA GRANDIFLORA 'BRAMDEAN'

A classy upright catmint bearing whorls of blue flowers, held in darker calices, on long stems. The foliage and stems darken in hot summers. From Victoria Wakefield's Bramdean garden in Hampshire. Height 75cm (2.5ft) / H5-9

'SUMMER NIGHTS' HELIOPSIS HELIANTHOIDES VAR. SCABRA

SALVIA 'AMISTAD'
This deep-blue salvia lights up any yellow companion, as do all blues. Not likely to overwinter for many, though. Height 1.2-1.5m (4-5ft) / H up to 8

HYLOTELEPHIUM 'JOSÉ AUBERGINE'
A dark and sumptuous sedum, with foliage that shares the same aubergine lustre. Compact and with deep-red flowerheads. Named by José De Buck, chairman of the Flemish Perennials Association. Height 45cm (18in) / H5-9

ACONITUM 'SPARK'S VARIETY'
A true-blue summer-flowering monkshood that adds rich blue highlights, rather than a solid spike. Highly toxic. Height 1.2m (4ft) / H4-8

EUTROCHIUM MACULATUM (ATROPURPUREUM GROUP) 'RIESENSCHIRM'
Previously listed as a eupatorium, this Joe Pye weed has black stems and purple-pink flowers. Needs summer moisture! Height 2m (6.5ft) / H3-8

AMSONIA TABERNAEMONTANA

GEUM 'TOTALLY TANGERINE'
Long-flowering, sterile geum with airy stems topped by frilly apricot flowers. The main flush coincides exactly with the blue starry amsonia flowers. Height 90cm (3ft) / H5-7

DICTAMNUS ALBUS 'BURNING BUSH'
Another slow-fuse affair, with strong upright stems and curious asymmetrical, beautifully veined flowers in purple or cream-white. Star-shaped seedpods follow. Slug-prone in early spring. Height 90cm (3ft) / H3-8

PENNISETUM ALOPECUROIDES 'CASSIAN'S CHOICE'
The brown fuzzy caterpillar heads mingle well with the yellowing foliage and dark seedheads. Needs good drainage. Height up to 90cm (3ft) / H5-9

TULIPA SPRENGERI
The last tulip to flower, with small burnished red goblets in May. Allow the seeds to drop and the brown seedheads, which dry beautifully, will endure into autumn. Height 30cm (12in) / H3-8

'SAHIN'S EARLY FLOWERER' HELENIUM

AGASTACHE 'BLUE FORTUNE'
Fuzzy, mid-blue bottlebrush heads stand erect above aromatic foliage. Height 90cm (3ft) / H4-10

SALVIA NEMOROSA 'CARADONNA'
A willowy, hardy perennial with black stems topped with dark-blue tapers. Mass plant this one. Height 50cm (20in) / H5-9

SALVIA NEMOROSA 'OSTFRIESLAND'

A front of the border salvia with violet flowers held in pink bracts – looking like a naked gas flame. Height 50cm (20in) / H5-9

NEPETA GRANDIFLORA 'SUMMER MAGIC'

A new catmint, with long-lasting pale-blue, bee-friendly flowers in whorls. This one won't flop. Height 70cm (28in) / H5-9

'ROMA' ASTRANTIA

HELLEBORUS X HYBRIDUS

Sultry pink and red astrantia, which flowers in May mostly, make good follow-up acts to spring-flowering hellebores. Height 30-35cm (12-14in) / H6-8

AQUILEGIA VULGARIS 'MUNSTEAD WHITE'

A stable seed strain, around in Gertrude Jekyll's day, with wide green-tipped white flowers and rounded green foliage. Best with deep-pink or red astrantias. Height 60cm (2ft) / H3-9

ROSA 'CHAMPAGNE MOMENT' (KORVANABER)

Astrantias do well in the lee of roses, pre-empting the main rose flush in June. This Kordes rose, sold all over the world but under different names, is a soft buff-cream. Rose of the Year UK 2006. Height 1.2m (4ft) / H5-9

CENTRANTHUS LECOQII

A hardy, Moroccan valerian with flowers that hover between pink and blue. Loved by hummingbird hawk-moths. Height 60cm (2ft) / H5-8

'TOTALLY TANGERINE' GEUM

AMSONIA TABERNAEMONTANA

Clusters of slate-blue flowers in May – the best partner of all. Height up to 90cm (3ft) / H4-9

HYLOTELEPHIUM TELEPHIUM (ATROPURPUREUM GROUP) 'KARFUNKELSTEIN'

A dusky-leaved sedum that has rosettes of pinkish-grey leaves at first, darkening to beetroot-red and then topped by substantial heads of red stonecrop flowers in August. Height up to 45cm (18in) / H4-9

CAREX TESTACEA

A swirl of brown fibre-optic-like filaments, emanating from a tight waist. It creates gentle movement among the apricot flowers. Height 40cm (16in) / H6-10

IRIS 'LANGPORT WREN'

Shades of purple and russet-brown highlight the paler geum flowers and the two should coincide because intermediate irises flower in the second half of May. The best of the Langport series, raised by Kelways of Somerset between 1968 and 1973. Height 65cm (26in)/ H4-9

'PURRSIAN BLUE' NEPETA X FAASSENII

ROSA 'BONICA'

A cluster-flowered pink rose that does well on poorer soil. Begins to flower in July and carries on until November. Height 1.2m (4ft) / H5-9

STIPA TENUISSIMA PONYTAIL GRASS

Wispy green 'hairs', which turn golden in summer, forever moving and swaying. Height 30cm (12in) / H7-11

ERIGERON KARVINSKIANUS

Small pink and white tiny daisies on low-growing plants. Self-seeds and the flowers darken to deep pink in hot summers. Height 20cm (8in) H7-10

HEMEROCALLIS 'PRIMAL SCREAM'

A pumpkin-orange daylily, introduced in 1994 by Curt Hanson of Ohio, and it's won lots of awards. Flowers in July for me. Height 75cm (2.5ft) / H3-9

'DUCHESSE DE NEMOURS' PAEONIA LACTIFLORA

TRITELIA LAXA 'QUEEN FABIOLA'

Blue bulb, resembling a subtle agapanthus, and forming papery seedheads. Best on a sunny edge. Height up to 60cm (2ft) / H6-10

GERANIUM PRATENSE 'MRS KENDALL CLARKE'

A finely-veined grey-blue meadow cranesbill. This will self-seed and not always truly. Height 90cm (3ft) / H3-8

CAMPANULA PERSICIFOLIA

The peach-leaved bellflower, with open blue or white bells. Good at popping up between peonies and roses. Height 90cm (3ft) / H3-8

ERYNGIUM BOURGATII

Summer-flowering blue thimbles, surrounded by starry bracts, held above variegated foliage. For the sunny edge. Height 75cm (2.5ft) / H5-8

'OSTFRIESLAND' SALVIA NEMOROSA

SALVIA OFFICINALIS 'BERGGARTEN'

A glaucous-leaved culinary sage with large round leaves and large blue-pink flowers. Needs space. Height 60cm (2ft) / H4-9

HYLOTELEPHIUM TELEPHIUM (ATROPURPUREUM GROUP) 'PURPLE EMPEROR'

Neat dark foliage, crimped around the edges, supports ruby-red flowers in the second half of summer. Compact. Graham Gough of Marchants Hardy Plants found it in his parents' garden. Height 60cm (2ft) / H4-9

STIPA TENUISSIMA PONYTAIL GRASS

The perfect foil, with swirling green filaments that turn straw-gold by midsummer. Height 30cm (12in) / H7-11

SCABIOSA OCHROLEUCA 'MOON DANCE'

A wiry-stemmed pale-yellow scabious with small button flowers that bob about above finely cut foliage. This is the compact version of the species. Height 30cm (12in) / H5-8

SPANISH OAT GRASS STIPA GIGANTEA

SYMPHYOTRICHUM NOVAE-ANGLIAE 'HELEN PICTON'

An easy New England aster with deep-purple flowers. The lower leaves tend to get ragged, so give it a middle of the border position please. Height up to 1.2m (4ft) / H4-8

SALVIA X SYLVESTRIS 'MAINACHT'

German May-flowering hybrid between *S. pratensis* and *S. nemorosa*, with deep-purple flowers, encased in reddish bracts. The earliest deep-blue salvia – but it splays out too much for some. Height 50cm (20in) / H5-9

BUDDLEJA DAVIDII 'BLACK KNIGHT'

This buddleia, with deep-purple honey-scented flowers and neat grey-green foliage, is enhanced by the golden rain produced by this tall stipa. Height 2.5m (8ft) / H5-10

ALLIUM X HOLLANDICUM 'PURPLE SENSATION'

The 'cheap-as-chips' go-to purple drumstick allium adds a touch of May drama to this stipa. Height up to 90cm (3ft) / H4-7

JAPANESE HAKONE GRASS HAKONECHLOA MACRA

RODGERSIA PINNATA

Bronze-tinged dark-green foliage followed by small pink flowers. Height 1.2m (4ft) / H5-8

KIRENGESHOMA PALMATA

Waxy yellow flowers above maple-leaf foliage. Needs lime-free soil and a large space to shine. From Japan and Korea. Height up to 1m (39in) / H4-9

DRYOPTERIS WALLICHIANA

Look for the black-bristled Himalayan form, which looks superb rising behind golden Hakone grasses. Height 70cm (28in) / H6-8

BERGENIA 'ERIC SMITH'

Bergenias thrive in sheltered areas that get some sun. The red-backed leaves of 'Eric Smith' turn deep purple-red in winter. Pink flowers follow. Height 45cm (18in) / H4-8

'PRICHARD'S VARIETY' CAMPANULA LACTIFLORA

KNAUTIA MACEDONICA

A maroon-red pincushion that begins to flower from May onwards. It does get ragged later in the season and it does not respond to shearing back. It also self-seeds. I still love it: it brightens up the fickle month of May while pleasing lots of pollinators. 'Knauty', but nice! Height 80cm (32in) / H5-9

HEMEROCALLIS 'CORKY'

Early, pale-yellow flowers, opening from mahogany-brown buds, clustered together on dark stems, held above narrow green foliage. Said to be from the same 1954 seedpod as the darker-yellow 'Golden Chimes'. Both are free-flowering. Fischer, 1959. Height up to 90cm (3ft) / H3-9

PHLOX PANICULATA 'MONICA LYNDEN-BELL'

A chalk-tolerant border phlox is a rare thing. This pale-pink phlox was named in the 1970s, or possibly 1980s, after the Hampshire gardener who found it in her own garden. 'Alba Grandiflora' also tolerates drier conditions. Height 70cm (28in) / H4-9

ROSA 'WILDEVE'

From David Austin, a healthy, wide-spreading rose with repeat clusters of pale-pink flowers in June. Height 1.2m (4ft) / H5-9

'SPARK'S VARIETY' ACONITUM

HELENIUM 'SAHIN'S EARLY FLOWERER'

Blue highlights this long-flowering orange helenium perfectly. Height 90cm (3ft) / H3-9

THALICTRUM DELAVAYI 'HEWITT'S DOUBLE'

A double meadow rue with tall sprays of tiny lilac double-flowers floating above ferny 'maidenhair' foliage. Height up to 1.5m (5ft) / H5-9

JARAVA ICHU OR STIPA ICHU

A flowing, clump-forming stipa with cream-white flowering heads by midsummer. Height 60cm (2ft), taller in good soil / H8-9

SCABIOSA OCHROLEUCA

Wiry stems topped with pallid-yellow pincushion flowers from the second half of summer onwards. Deadhead! Height 90cm (3ft) / H5-8

'MÖNCH' ASTER X FRIKARTII

RUDBECKIA FULGIDA VAR. DEAMII
This crisp brown-centred yellow daisy has more upright stems topped by larger daisies. Height 90cm (3ft) / H3-10

PENSTEMON 'ANDENKEN AN FRIEDRICH HAHN'
Another Swiss-bred golden oldie, raised in the 1920s. Lots of slender wine-red flowers above narrow foliage. Hardier than many. Height 90cm (3ft) / H7-10

HYLOTELEPHIUM 'MATRONA'
A sturdy sedum with fleshy pigeon-grey foliage and dusky-pink flowers. Height 90cm (3ft) / H3-9

GERANIUM 'DRAGONHEART'
Black-eyed magenta flowers for months on end, plus good foliage. Height 60cm (2ft) / H3-8

'GEORGE DAVISON' CROCOSMIA

KNIPHOFIA 'SAFRANVOGEL'
Short poker bearing masses of small peach-salmon to orange-red torches in midsummer. Height 80cm (32in) / H5-9

SCABIOSA COLUMBARIA
Late-summer, small-flowered powder-blue scabious, from wiry stems, underpinned by highly divided foliage. Easy from seed. Height 60cm (2ft) / H5-8

CLEMATIS HERACLEIFOLIA 'CASSANDRA'
Herbaceous (i.e. non-climbing) late-summer clematis with whorls of deep-blue fragrant flowers – a bee and butterfly magnet. Height 90cm (3ft) / H4-9

NEPETA 'CHETTLE BLUE'
An aromatic, upright catmint with dazzling blue flowers held in darker calices. Flowers forever. Height 90cm (3ft) / H5-9

'GLOBEMASTER' ALLIUM

STIPA GIGANTEA
A golden fountain that arrives with the alliums. Height up to 2m (6.5ft) / H5-10

ASTRANTIA 'ROMA'
Candy-pink. Long-flowering crisp astrantia. Height 50cm (20in) / H4-9

ANTHEMIS 'POWIS CASTLE'
Grown for its aromatic silver foliage, this wormwood hides faded allium foliage. Height 60cm (2ft) / H4-8

ANTHEMIS 'SUSANNA MITCHELL'
A floppy anthemis with fine grey-green foliage and pale-yellow daisies. Height 45cm (18in) H3-7

'MONICA LYNDEN-BELL' PHLOX PANICULATA

ECHINACEA PURPUREA
Stiff stems, topped by pink or white petals set around a bronzed middle, add contrast. Height on average 90cm (3ft) / H3–9

PAEONIA LACTIFLORA
These June-flowering peonies come before the phloxes, but both enjoy similar conditions. Height 1m (39in) / H3–8

VIOLA CORNUTA
The winged violet is truly perennial, with a mat of small leaves below long-stemmed May flowers. Height 15cm (6in) / H6–9

MONARDA 'VIOLET QUEEN'
A strong performer, with heads of punk-like purple flowers held on strong stems. Use with the pinks and orange-pink phloxes. Height 90cm (3ft) / H4–9

ERYNGIUM X ZABELII

HEMEROCALLIS 'HYPERION'
A golden-oldie from 1924, but (just like 'Whichford') a great performer, with lots of pale-yellow flowers. Height 90cm (3ft) / H3–9

ACHILLEA 'TERRACOTTA'
A sun-loving yarrow with curly grey-green foliage and terracotta flowers that fade to paper-brown. Height 80cm (32in) / H3–8

GEUM 'SCARLET TEMPEST'
Not really scarlet, but redder than sun-ripened apricots, with early summer flowers that aren't too large or threatening. Needs some moisture. From Elizabeth MacGregor's nursery in Kirkcudbright. Deadhead this one. Height up to 80cm (32in) / H5–9

STIPA TENUISSIMA
Allow the straw-like filaments to curl around the eryngium flowers. Height 30cm (12in) H7–11

'JACOB CLINE' MONARDA

HYLOTELEPHIUM 'JOSÉ AUBERGINE'
The dark foliage of this sedum matches the soot-stained bracts under the red flowers. Height 45cm (18in) / H5–9

ACTEA SIMPLEX 'PINK SPIKE'
Bronze-purple foliage and fine tapers of pale-pink flowers – needs some shade. Height 1.2m (4ft) / H4–9

DAHLIA 'BISHOP OF LLANDAFF'
The classic peony-flowered red dahlia, named in the 1920s, backed by elaborate ferny dark foliage.It's another smouldering blaze of fire. Height 1.2m (4ft) / H8–11

CROCOSMIA 'COLUMBUS'
A pink-tinged herringbone of buds opens to warm-yellow on this neat, not-too-tall crocosmia. Height 90cm (3ft) / H6–9

'HERRENHAUSEN' ORIGANUM LAEVIGATUM

SESELI HIPPOMARATHRUM

Feathery evergreen foliage and small moody-pink umbels of flower in August, held on long stems. Height 90cm (3ft) at most / H5–9

POTENTILLA NEPALENSIS 'RON MCBEATH'

Small dark-centred cherry-pink flowers with a hint of red, are held on long stems. From midsummer onwards. Height 90cm (3ft) / H5–9

NEPETA RACEMOSA 'AMELIA'

The only pink-flowered catmint to come through British winters successfully. Forms a loose mound in summer. Height 60cm (2ft) / H5–9

CALAMINTHA NEPETA SUBSP. NEPETA 'BLUE CLOUD'

Pale-blue flowers flecked through menthol-scented foliage. Can self-seed too enthusiastically for some. Height 40cm (16in) / H5–9

'ANDENKEN AN FRIEDRICH HAHN' PENSTEMON

LAVANDULA ANGUSTIFOLIA 'MUNSTEAD'

Paler mauve English lavender that usually overlaps with this wine-red penstemon. Height 50cm (20in) / H5–9

NEPETA GRANDIFLORA 'BRAMDEAN'

Upright stems of blue flowers held in darker calices. Height 75cm (2.5ft) / H5–9

PHLOX PANICULATA 'FRANZ SCHUBERT'

Alan Bloom's homage to his favourite composer, with pale flowers that hover between mauve and blue. Height 90cm (3ft) / H4–9

SCABIOSA CAUCASICA

Frilly large blue flowers held on stems that bow their necks. A brilliant cut flower. Height 90cm (3ft) / H4–9

VAR. DEAMII RUDBECKIA FULGIDA

ASTER X FRIKARTII 'MÖNCH'

Long-lashed lavender-blue daisies that go on just as long. Height 90cm (3ft) / H4–9

SUCCISELLA INFLEXA 'FROSTED PEARLS'

This scabious relative has small lavender bobbles from late summer onwards. Very pollinator-friendly. Height 90cm (3ft) / H4–9

SYMPHYOTRICHUM LAEVE 'LITTLE CARLOW'

A cloud of tiny rich-blue flowers, held in red-patterned buds, is a fine backdrop. Height 1.2m (4ft) / H4–9

PHLOX PANICULATA 'USPEKH'

This Russian border phlox is later flowering than most, with white-eyed purple flowers on a shorter plant. Height up to 90cm (3ft) / H4–9

'MATRONA' HYLOTELEPHIUM

ORIGANUM 'ROSENKUPPEL'

A hybrid origanum with airy heads of purple-pink flowers over a long period. Height 60cm (2ft) / H5 - 9

STACHYS OFFICINALIS 'HUMMELO'

A vibrant betony with heads of pink-purple flowers. Height 70cm (28in) / H4-9

NEPETA RACEMOSA 'WALKER'S LOW'

A soft haze of blue flowers, held above small grey-green leaves. Less floppy and not that low. Height 60cm (2ft) / H5-9

GERANIUM 'PATRICIA'

Magenta flowers, middled in black, from August onwards, on this large, billowing, hardy geranium. Height up to 1.2m (4ft) / H4-8

'ALBUM' VERBASCUM CHAIXII

ANEMONE HUPEHENSIS VAR. JAPONICA 'PAMINA'

Bright-pink semi-double flowers, emerging from grey seed-pearl buds on this dark-stemmed Japanese anemone. It runs and fills gaps. Height 75cm (2.5ft) / H4-8

SIDALCEA 'WILLIAM SMITH'

A startling deeper-pink sidalcea which will pick up the dusky blueberry splash on this white verbascum. Height 90cm (3ft) / H4-9

LIMONIUM PLATYPHYLLUM SEA LAVENDER

Plant this sea lavender, grown for its purple froth of August flowers, on the sunny side of your border. Height 90cm (3ft) / H4-8

VERBENA 'BAMPTON'

Dark foliage on this bush verbena. The small mauve flowers go on until really late. Self-seeds – but not aggressively so. Height up to 90cm (3ft) H4-8

'ORION' GERANIUM

ROSA 'CHAMPAGNE MOMENT'

The buff-white flowers of this extremely healthy floribunda rose enhance the true-blue flowers. Height 1.2m (4ft) / H5-9

STACHYS BYZANTINA

Felted silver foliage and spires of soft-pink flowers on short stems. Height 30cm (12in) / H4-9

KNAUTIA MACEDONICA

A deep-red scabious that begins to flower in May, just as 'Orion' produces its first flowers. Deadhead this one. Height 80cm (32in) / H5-9

ALCHEMILLA MOLLIS

Lady's mantle, another notorious self-seeding perennial, I admit. However, the fluffy lime-yellow heads set off the blue saucers brilliantly well. Height 60cm (2ft) / H5-10

'SEAL' LAVANDULA

PHLOMIS 'EDWARD BOWLES'

A small subshrub with large downy heart-shaped leaves and whorls of sulphur-yellow flowers in summer. A good plant for a sunny bank in free-draining soil. A compact hybrid between *P. fruiticosa* and the spreading *P. russeliana*. Height 1m (39in) / H7-10

SANTOLINA CHAMAECYPARISSUS SYN. S. INCANA

Forms round dense bushes of silver-grey, closely feathered foliage and small yellow button flowers in summer. Clipped in September, it makes another winter roundel. Height up to 50cm (20in) / H5-10

HYLOTELEPHIUM TELEPHIUM 'PURPLE EMPEROR'

Dark foliage from the moment it emerges in spring, with neatly crimped edges to the leaves, followed by ruby-red flowers in August. A neat sedum. Height 60cm (2ft) / H4-9

SILENE BANKSIA 'HILL GROUNDS'

A silver-leaved campion with bright-pink flowers in summer. It's shorter and does not self-seed. From the late Janet Cropley's garden in Northamptonshire. Height 60cm (2ft) / H4-9

'HONORINE JOBERT' JAPANESE ANEMONE
ANEMONE X HYBRIDA

FUCHSIA 'MRS POPPLE'

Japanese anemones are good at popping up around hardier fuchsias, in a shady strip by a path. 'Mrs Popple' is a sumptuous red and purple fuchsia from the 1930s, via Clarence Elliott. Height 1.2m (4ft) / H8-12

CENTRANTHUS LECOQII

A Moroccan valerian with mauve flowers, good with all pinks. Hardy too. Height 60cm (2ft) / H5-8

ROSA 'ROSY CUSHION'

A long-flowering modern shrub rose, from 1979, with clusters of single, scented flowers. Healthy and hardly ever out of flower. Height up to 1.5m (5ft) / H5-9

CHRYSANTHEMUM 'CLARA CURTIS'

A single-pink hardy chrysanthemum that flowers prolifically in September. Needs good drainage. Plant with *Nerine bowdenii*. Height 80cm (32in) / H3-9

'FERNER OSTEN' MISCANTHUS SINENSIS

SANGUISORBA 'CANGSHAN CRANBERRY'

A tall red-bobbled perennial that flowers in late September. Less willowy than many. Height 2m (6.5ft) / H4-8

ASTER LAEVIS 'CALLIOPE'

Dark, almost black leaves and stems and large lavender-blue flowers in September. Height 1.2m (4ft) / H4-9

VERNONIA ARKANSANA (PREVIOUSLY V. CRINITA)

A North American aster lookalike with tight heads of purple flowers – also in September. Height 1.5m (5ft) / H5-8

HELIANTHUS 'LEMON QUEEN'

A dark-centred lemon daisy and one of the few that will stay in a tight clump. Height 1.5m (5ft) / H3-9

'ARENDSII' ACONITUM CARMICHAELII

ROSA 'SWEET HONEY'

This healthy Rose of the Year 2020, raised by Kordes, has clusters of perfectly formed small flowers that open peach and fade to clotted cream. Use with shorter sanguisorbas. Height 90cm (3ft) / H5-9

MISCANTHUS SINENSIS 'ZEBRINUS'

The yellow-barred green foliage sets off the dark bobbles. Height 1.5m (5ft) / H4-9

ECHINACEA PURPUREA 'WHITE SWAN'

Seed-raised white coneflower, with a burnished golden middle, throws up the light. Height 80cm (32in) / H3-9

SALVIA 'BLUE SPIRE' SYN. PEROVSKIA

White branching stems clothed in small lavender-blue flowers create a haze. Height up to 1.2m (4ft) / H4-9

'LITTLE CARLOW' SYMPHYOTRICHUM

HESPERANTHA COCCCINEA 'MAJOR'

Large, coppery-red saucers above grassy foliage on this moisture-loving South African plant. Height 60cm (2ft) / H6-10

CROCOSMIA 'LUCIFER'

By the time 'Little Carlow' flowers, this substantial crocosmia will only be showing green, upright sword-shaped foliage. Height up to 1.5m (5ft) / H5-9

SOLIDAGO RUGOSA 'LOYDSER CROWN'

Large arching sprays of sherbet-lemon flowers. This drought-tolerant cultivar makes a good cut flower and insects love it. Height 1.2m (4ft) / H3-9

RUDBECKIA SUBTOMENTOSA 'LOOFAHSA WHEATEN GOLD'

Large yellow daisies with swooning petals. Bred by Anthony Brooks of Elton Hall and named after his terrier dogs. Prefers some moisture. Height 1.5m (5ft) / H3-10

'HERBSTSONNE' (AUTUMN SUN) RUDBECKIA LACINIATA

CLEMATIS 'PURPUREA PLENA ELEGANS'

A double damson-purple viticella clematis that could be mistaken for a small rosebud. This will happily scramble through the stems of tall, upright daisies. Height 1.5m (5ft) / H8–11

CORTADERIA RICHARDII

New Zealand's own pampas grass, the Toe Ioe. It produces flowing feathery heads from July onwards. The foliage is a little savage, so place it carefully. Height up to 3m (10ft) / H7–8

ALLIUM 'SUMMER DRUMMER'

A flowering leek, producing leaves in winter, and then a tall rubbery stem topped by a dusky-pink head in summer, roughly the size of a tennis ball. The heads darken as autumn descends. Introduced in 2006 by W. Mellema. Best planted in groups of five or seven. Height 2m (6.5ft) / H5–9

CALAMAGROSTIS X ACUTIFLORA 'KARL FOERSTER'

Slender feathery heads on a sheath of stems. This is the most upright grass of all. Height 1.5m (5ft) / H3–9

(ATROPURPUREUM GROUP) EUTROCHIUM MACULATUM

HELENIUM AUTUMNALE 'FLAMMENDES KÄTHCHEN' – FLAMING KATIE

This red-orange helenium flowers later, because of the height it has to reach and that takes time. Height up to 1.5m (5ft) / H3–9

PHLOX X ARENDSII 'HESPERIS'

Upright sturdy stems carry domed heads of scented lilac-pink flowers. These glow in evening light. Moths love it. Height 90cm (3ft) / H3–9

SANGUISORBA OFFICINALIS 'MARTIN'S MULBERRY'

The deep-red bobbles pick up the clouds of dusky flowers. Place a distance away though, to link the two with the eye. Height 1.8m (6ft) / H4–8

VERONICASTRUM VIRGINICUM 'FASCINATION'

The blue flower spikes flatten to produce green-tipped mermaid's tails and I have wondered whether the name is a play on fasciation. Height up to 1.5m (5ft) / H3–9

CREDITS

The publishers would like to thank the following sources for their kind permission to reproduce the pictures in this book.

Alamy Stock Photo: RM Floral 46-47; /Anne Gilbert 90-91; /Martin Hughes-Jones 112-113; /imageBROKER.com GmBH & Co. KG 6-7; /Rex May 48-49; /John Richmond 104-105; /Matthew Taylor 108-109; /Deborah Vernon 64-65

Gap Gardens: Richard Bloom 36-37; /Marg Cousens 34-35; /Ray Cox 86-87; /Martin Hughes-Jones78-79; /Lynn Keddie 32-33; /Howard Rice 96-97; /Visions 58-59

© MMGI/Marianne Majerus: 12-13, 38-39, 40-41, 42-43, 44-45, 50-51, 52-53, 54-55, 56-57, 60-61, 62-63, 66-67, 68-69, 70-71, 72-73, 74-75, 80-81, 82-83, 84-85, 88-89, 92-93, 94-95, 98-99, 100-101, 102-103, 106-107; /© MMGI/Bennet Smith: 19

Val Bourne: 14-15, 16-17, 20-21, 26-27

Shutterstock: Egschiller 24-25; /Alex Manders 76-77; /Andrey Nikitin 110-111

ACKNOWLEDGEMENTS

My thanks to the team at Welbeck: my publisher Heather Boisseau for her encouragement, Matt Tomlinson for his editorial work, and Chris Stone for editing the book.
Thanks also to the 'Best Beloved', Jo Kirby, for his patience and sandwich-making skills.